# EXTENDING SCIENCE 7

## ENERGY
## Selected Topics

Extending Science Series

| | | |
|---|---|---|
| 1 | Air | E N Ramsden and R E Lee |
| 2 | Water | E N Ramsden and R E Lee |
| 3 | Diseases and Disorders | P T Bunyan |
| 4 | Sounds | J J Wellington |
| 5 | Metals and Alloys | E N Ramsden |
| 6 | Land and Soil | R E Lee |
| 7 | Energy | J J Wellington |

Further titles are being planned, and the publishers would be grateful for suggestions from teachers.

# EXTENDING SCIENCE 7

# Energy

## Selected Topics

J J Wellington BSc MA

Stanley Thornes (Publishers) Ltd

First published 1985 by
Stanley Thornes (Publishers) Ltd
Old Station Drive
Leckhampton
CHELTENHAM GL53 0DN



British Library Cataloguing in Publication Data

Wellington, J.J.
Energy. — (Extending science; no. 7)
1. Power resources
I. Title II. Series
333.79 TJ163.2

ISBN 0-85950-237-6

Typeset by Tech-Set, 15 Enterprise House, Team Valley, Tyne & Wear.
Printed and bound in Great Britain by the Bath Press Ltd., Bath.

# CONTENTS

## Chapter 7 **Energy and Different Life Styles**

# PREFACE

Your life depends on energy. You need energy to move, to travel to work or school, to play games, to keep warm or just to stay alive.

Energy is one of the most important topics in science.

The following pages set out to explain:

- what energy is
- how energy is converted from one form to another
- where energy comes from
- how energy is measured
- how you receive and then convert your energy
- why energy supplies are a problem
- how energy can be saved
- why people talk about an energy crisis

J J Wellington

# ACKNOWLEDGEMENTS

Barnaby's Picture Library (pp. 20, 40, 50, 58, 59 (a), (c) and (d))
Alan Beaumont (p. 4 left)
Paul Brierly (p. 24)
British Petroleum (p. 4 right)
British Rail, Western (pp. 59 (b), 61)
Central Electricity Generating Board (pp. 12, 34, 35, 37)
Digital Equipment Co Ltd/Wimpey Laboratories Limited (p. 39)
NASA (p. 62)
National Centre for Alternative Technology (p. 36)
National Coal Board (p. 23)
Syndication International (p. 57)
Tate and Lyle Group Research and Development (p. 41)
United Kingdom Atomic Energy Authority (p. 30)
*Cover*
All-Sport Photographic Agency Limited and The Central Electricity Generating Board

*I would like to thank my wife, Wendy, for help at every stage of the manuscript, proofs and index.*

J J Wellington

# CHAPTER 1

# THE MEANING OF ENERGY

## COMMON IDEAS ABOUT ENERGY

Most people have an idea of what energy is. If you ask people, they may say things like this:

- it's coal, oil and gas
- energy makes things go
- you need energy to play football
- everyone needs energy to live
- you get energy from food
- exercise uses up your energy
- energy gets things done
- energy keeps us warm
- electricity has energy

All these statements and ideas about energy are true, in their own way. But in science we need a more exact meaning for 'energy'.

You can't see energy — but it does make things happen!

## WHAT IS ENERGY?

The problem is that you cannot *see energy.* You can see things or people which have energy; you can see things happening because of energy; you can see things changing because of energy, e.g. growing. But you don't see energy itself.

*Energy makes things happen*
*Energy makes things change*

Here are some examples of changes which involve energy:

- a box being lifted
- an athlete running
- a fire burning
- an electric fire glowing
- a plant growing
- an aeroplane taking off
- a bulb lighting a room
- a bomb exploding

Each of these changes involves energy. You can understand this more easily when you realise that there are several different forms of energy.

## DIFFERENT FORMS OF ENERGY

### Kinetic energy

The wind has energy. A waterfall has energy. A moving rocket has energy. A bird flying has energy. They all have energy because they are moving. They have *movement energy,* which is usually called *kinetic energy.*

### Potential energy

A tensed muscle has energy. A raised weight has energy. A compressed spring has energy. This energy is *stored,* waiting to be converted. It is called *potential energy.* If the spring is released or the weight is dropped a *change* takes place, e.g. the weight picks up speed.

### Chemical energy

Energy is stored in fuels, such as gas, coal, oil, or wood. Energy is stored in food. This stored energy is called *chemical energy.* The energy stored in a battery is chemical energy. This energy can make things change.

### Wave energy

The Sun's rays reach to the Earth. These rays have energy. They can bring about changes — they can warm you up, light up a room, help plants to grow, or even give you a suntan. The Sun's energy is carried by waves: light waves, infra-red waves and ultra-violet waves. This can be called *wave energy.* Sound waves, too, carry wave energy — an ultrasonic wave can bore a hole in wood. Water waves carry energy.

### Electrical energy

Electricity, like a wave, carries energy from one place to another, e.g. from a power station to your home. This energy can be used to make changes — to drive electrical motors, warm a room, operate your television set and so on. It is called *electrical energy.* Electrical energy is often the most convenient form of energy.

### Thermal energy

Heat, or *thermal energy,* raises the temperature of things, e.g. of cold water, your body, a furnace. So thermal energy brings about useful changes. It can be used to drive turbines, melt steel, or simply to boil water. (You will see later that all forms of energy *end up* as heat. But this heat is at a low temperature and is of no use to us.)

### Nuclear energy

Nuclear energy is sometimes called *atomic energy.* It is released when small atoms are forced together, or when large atoms are split apart. Nuclear energy is released carefully in power stations, or in an uncontrolled way when a nuclear bomb explodes. Most of the energy released is in fact wave energy and thermal energy, but energy from the atom is given its own name: nuclear energy. We shall learn later that the Sun's energy is nuclear energy.

All these names are given to the different types of energy that you see around you. These forms of energy all have *one* thing in common: they can all bring about changes, or make things happen. You will see many examples of this in later pages.

## WHERE DOES ENERGY COME FROM?

Energy is needed for transport, home heating, cooking and simply staying alive. This energy may come from petrol, gas, oil, coal and so on. These are all called *sources of energy.*

### Capital

The most common sources of energy are *fuels.* Nearly half of Britain's energy comes from one fuel alone, oil. Most of the energy we use comes from inside the Earth itself. Oil, gas and coal all have stored, chemical energy, which took millions of years to develop under our feet. This stored energy is our energy *capital.* It is there waiting to be used. But once used it has gone for ever.

The windmill uses energy income. The oil rig uses energy capital

### Income

The other type of energy source is present all the time, above and below the Earth. Unlike fuels (our energy capital) it is constantly 'coming in'. This is energy from the wind, the waves, daily tides and the Sun's radiant energy. It is called energy *income.* The Sun's energy is supplied continuously. So is the energy of the wind and of the tides. These energy sources are not used up. They are often called *renewable sources.*

## BALANCING THE 'ENERGY BOOKS'

Coal, oil and gas (our energy capital) were formed millions of years ago. Yet within fifty years from now, most of the world's known oil and gas will have been used up. In other words, our energy capital is being *used up* much faster than it is being replaced. This is like having a bank account with money (capital) in it which you keep taking out. Eventually your account will be empty.

The energy balance: supply and demand

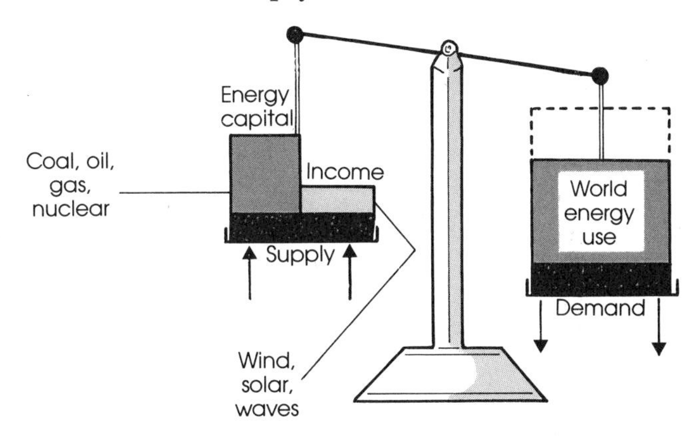


There are two solutions: to take less out, and to put more in. In terms of energy this means three things:

- new sources of energy *capital* are needed
- the world must use less energy
- our energy *income* must be used properly

These are the three important themes in the rest of this book . . .

## QUESTIONS ON CHAPTER 1

**1** Supply words or phrases to fill the blanks in the following question. Do not write on this page.

(a) A moving athlete has ____ energy.
(b) Fuels like oil and gas contain ____ energy.
(c) A stretched spring possesses ____ energy.
(d) The energy which reaches us from the sun can be called ____ energy.
(e) Sound is one type of ____ energy.
(f) The two types of energy source are called energy ______ and energy ______.

**2** Imagine that somebody asks you what energy is. Explain, in your own words, what you would say.

**3** Look at the list of the different forms of energy. Make a simple table showing each form, with two everyday examples of each and a simple drawing to illustrate it, e.g.

*Kinetic energy*
bird flying
moving car

**4** Name an energy source or energy sources for each of the following:

(a) your personal energy

(b) the energy you use for *travelling around*

(c) the energy you use for *keeping warm*

(d) the energy you use for *cooking your meals.*

**5** Look at the picture on p. 1. Discuss the 'energy situations'. Can you see evidence for all seven types of energy?

## GLOBAL ENERGY WORDFINDER

First, trace the wordfinder onto a piece of paper (or photocopy this page — teacher, please see the note at the front of the book). Then find ten energy sources and list them in two columns labelled *Renewable* and *Non-renewable.* The unused letters taken in order, row by row, will spell out a question that is worrying a lot of people.

CHAPTER 2

# ENERGY CONVERSIONS

## ONE FORM INTO ANOTHER

You have seen the names given to different forms of energy. Energy is constantly being changed from *one* of these forms into *another.* Indeed you only *notice* energy when it is being changed from one form into another.

The picture below shows three ways in which energy is changed from one form to another. Whenever this happens people say that energy is being 'used'. Here are some examples of common energy conversions:

Three examples of energy conversion

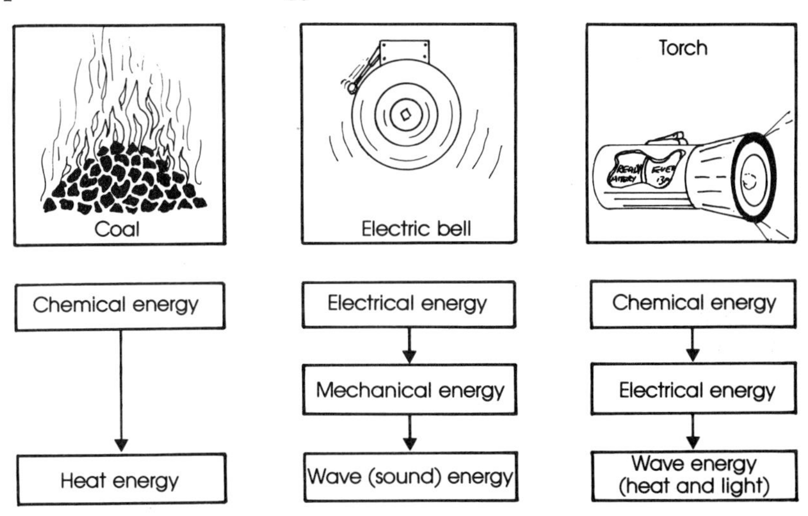


These are all energy conversions. Plants and animals, engines and motors, are all involved in energy conversions. They can all be called *energy converters.*

## ENERGY CONVERTERS

Plants are the most important energy converters on the Earth. They change the Sun's wave energy (radiant heat and light) into *chemical energy.* This chemical energy is stored in the plant. These energy changes are shown in the next illustration.

**Plants are energy converters**

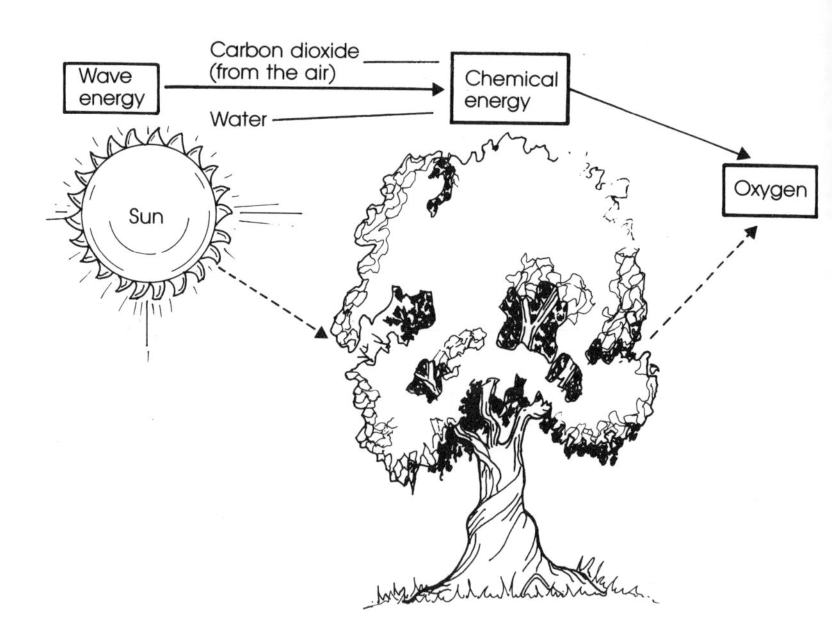


Plants need carbon dioxide from the air and water to make this energy change. The energy conversion is called *photosynthesis.* This means 'making something using light'. Photosynthesis takes place in the green parts of a plant. Plants *take in* carbon dioxide and *give out* oxygen. They provide animals with two things which keep them alive: *food* and *oxygen.*

Plants provide humans and animals with energy which they use to move around, to help them grow and to keep them warm. The figure below shows the energy changes involved. (Notice that heat energy is at the end of the 'energy chain' — you will see that this is true of most energy changes.)

**Humans are energy converters**

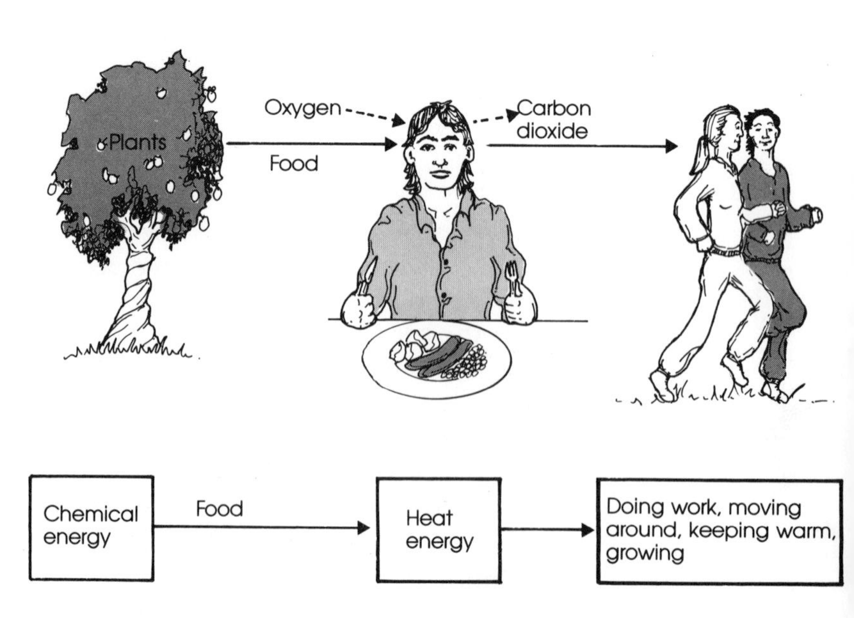


| Chemical energy | Food → | Heat energy | → | Doing work, moving around, keeping warm, growing |
|---|---|---|---|---|

Plants and animals are two types of *living* energy converters. But people have invented many *non-living* energy converters, which they can use to bring about useful changes and do useful jobs for them. Non-living energy converters can produce useful results, such as:

- cooking meals
- launching rockets
- heating homes
- boiling water
- lighting buildings
- generating electricity

ACTIVITY 1

## Listing energy converters

Make a list of energy converters for each of the six examples of energy conversion above.

The next section describes the most useful energy converters ever devised: *engines.* You will also find out some of their drawbacks.

## Engines

In 1860 a man from Luxembourg called Jean Lenoir invented a new type of engine. It needed a mixture of air and gas which exploded inside a metal cylinder. This explosion forced a piston to move up and down.

Lenoir was never honoured or rewarded for his invention. But it was in fact *the first internal combustion engine.* Nowadays the same kind of engine is used in cars, lorries, planes, machines and trains all over the world. It has totally changed the style of people's lives.

## The car engine

The car engine converts chemical energy from petrol into kinetic energy. The engine has a container or cylinder with a piston inside it. The cylinder has two openings controlled by valves. As the piston moves down, the inlet valve opens and sucks petrol and air in. Both openings are then closed by the two valves and the piston moves up, squeezing or compressing the mixture of petrol and air. Suddenly a spark from the sparking plug makes the petrol explode at just the right moment. The piston is pushed back down with a terrific force. When the piston moves up again it pushes out the used or burnt petrol through the second opening, the exhaust outlet. For every explosion the piston moves up and down twice. These are called the four strokes of the petrol engine: the intake of petrol and air, compression, explosion, and exhaust.

The car engine, an example of internal combustion

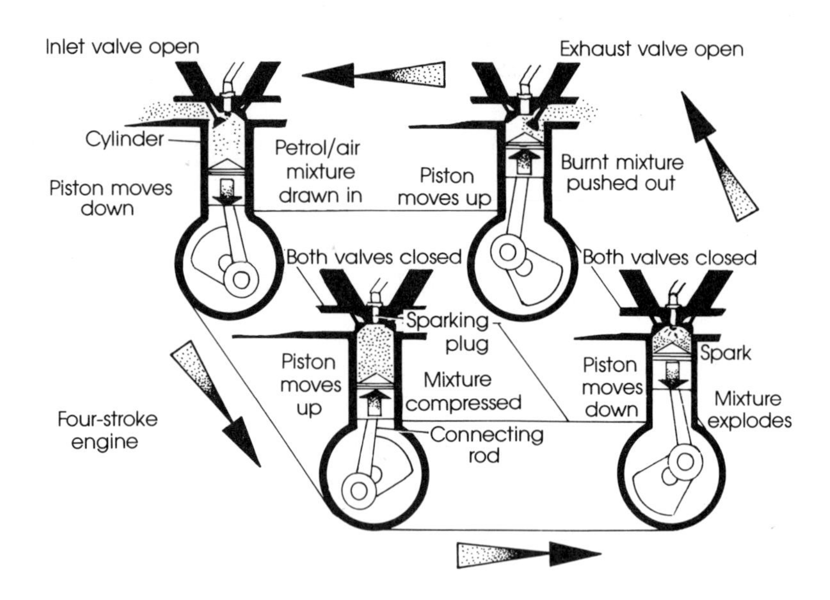


The way a car engine works is an example of *internal combustion.* Internal combustion engines mix a fuel with air, then burn this mixture *inside* the engine. Another example of an internal combustion engine is the *diesel* engine, invented by Rudolph Diesel in 1892. This uses diesel fuel instead of petrol, but both fuels are made from oil. Diesels are used in most lorries and taxis, in some cars, and for many train engines.

## Jets and rockets

Jet and rocket engines are two other important engines. They too convert chemical energy (from fuel) into kinetic energy and potential energy. The jet engine works by burning a mixture of fuel and air. This forms a mixture of very hot gases which rushes out of the back of the engine. As the gases rush backwards, the engine is pushed forwards.

Rockets work in a similar way. However, they need to travel in space where there is no air, so they carry their own supply of liquid oxygen. This oxygen mixes with the rocket fuel and burns in a special combustion chamber. The hot, fast moving gases are forced out of the back of the rocket through a nozzle. As the gases move backwards the rocket is pushed forwards.

The jet engine

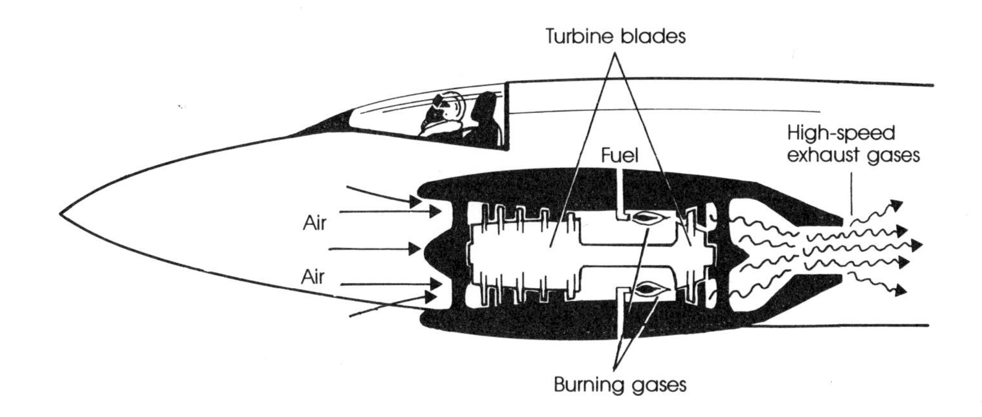


## The trouble with engines

The steam engine, car, rocket and jet engines have totally changed our lives. Engines of different sorts can be used to drive machinery, to carry goods, to transport people from one part of the Earth to another, or even to launch satellites into space. They may seem like the answer to all our problems. But engines have two big disadvantages:

- They all need a *fuel* of some kind, e.g. coal, petrol, diesel, rocket fuel. This fuel is part of the Earth's energy capital. The number of engines is going up all the time. The amount of fuel capital is becoming less and less. Eventually these fuels may run out (*see later*).
- Every engine wastes a lot of energy. *Most* of the chemical energy they convert is *wasted* as thermal energy (heat), which escapes into the air instead of being converted into kinetic energy. For example, in most car engines about three-quarters (75%) of the chemical energy from petrol is converted into heat which simply escapes into the air. Only a quarter (25%) of the chemical energy actually goes towards making the car move.

Engineers are trying all the time to make engines less wasteful or more *efficient.* But even the rocket engine is only 50% efficient, i.e. only *half* of the chemical energy from its fuel is converted into useful energy (kinetic and potential).

## THE LAWS OF ENERGY CONVERSIONS

The way that energy changes from one form into another is ruled by two important scientific laws. One is concerned with *quantity,* the other with *quality.*

### The law of quantity

This is sometimes called the law of conservation of energy. It states that *whenever energy is converted from one form into another, the total quantity of energy remains the same.*

In other words, energy is not destroyed but *conserved.* 'Excellent!' you may think. 'The world will never run out of energy, because it is always conserved.' So why do people worry about an energy crisis? This is where the second law comes in . . .

### The law of quality

The quantity of energy in the universe may stay the same — but the quality (unfortunately) gets worse: *whenever energy is converted from one form into another the overall quality of the energy gets worse.*

At the end of every energy conversion heat is produced. This is *low-temperature* heat energy which is difficult to use again. It simply warms up the Earth's atmosphere by a tiny amount. We say that this type of energy is of a 'poor quality'.

So after each energy change less usable energy remains. Here are three simple examples:

- Chemical energy in petrol is used to propel a car. Most of the energy ends up as low-temperature heat in the Earth's atmosphere.
- A torch battery converts chemical into electrical energy to light a torch. This energy ends up as light and heat energy which warms the air around the torch. The energy is still there — but it cannot be used again.

Coal-fired power stations in Staffordshire

- Some power stations convert the chemical energy from coal into electrical energy. Much of this energy becomes heat energy at the power station and escapes into the air through cooling towers. And finally, when people use electricity at home most of it is converted into heat by electric cookers, kettles or fires. This heat energy cannot be used again.

## HIGH-QUALITY AND LOW-QUALITY ENERGY

Although energy is not destroyed, its quality gets worse with every energy conversion. Examples of *high-quality* energy are:

- chemical energy in most fuels, e.g. petrol, coal, oil
- electrical energy
- high-temperature heat energy.

The best example of *low-quality* energy is low-temperature heat energy. This lies at the end of every energy conversion and is often useless.

**ACTIVITY 2**

### Energy circus

Try each of these energy conversions:

bell ringing; battery lighting a bulb; heating by friction (rubbing); magnesium ribbon burning; spring oscillating; match igniting; pendulum swinging; clockwork toy; thermocouple; inverting lead shot — i.e. ten examples (see illustration below).

In each case:

(a) What form of energy did you begin with?
(b) What form of energy was this converted to?

A circus of energy conversions

**ACTIVITY 3**

## Making a model rocket

You can see how a rocket works just by blowing up a balloon and letting it go. But you can make a better model using: a long thin balloon, a cardboard tube (e.g. the inside of a toilet roll), string and Sellotape. Stick two loops of Sellotape onto the cardboard tube. Thread a long piece of string through the tube and tie it to a door handle. Blow up the long balloon and hold the end. Then stick it on the tube and let go!

Making a model rocket

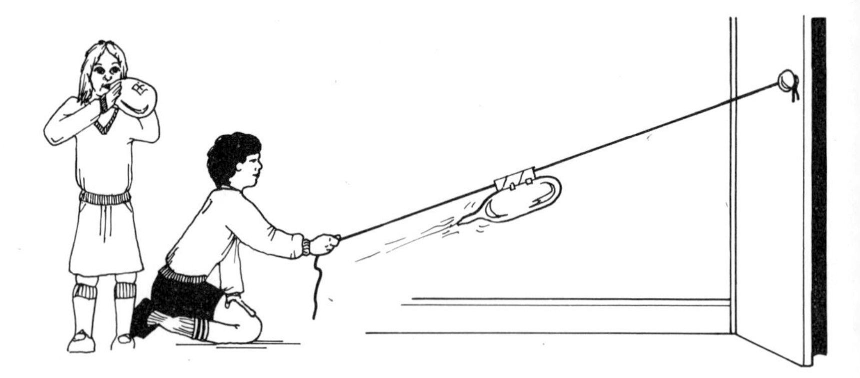

(a) What happens?
(b) Explain how it works.
(c) What forms of energy are involved in this model? (See illustration above.)

**ACTIVITY 4**

## Making a wind-up tank

You can make a model tank which moves slowly forward using: an old cotton reel, a candle, three matchsticks, a knife and a small rubber band:

1) Cut a thin slice off the end of the candle.
2) Bore a hole in the centre of this slice and cut a groove in it (see the illustration below).

Making a wind-up tank

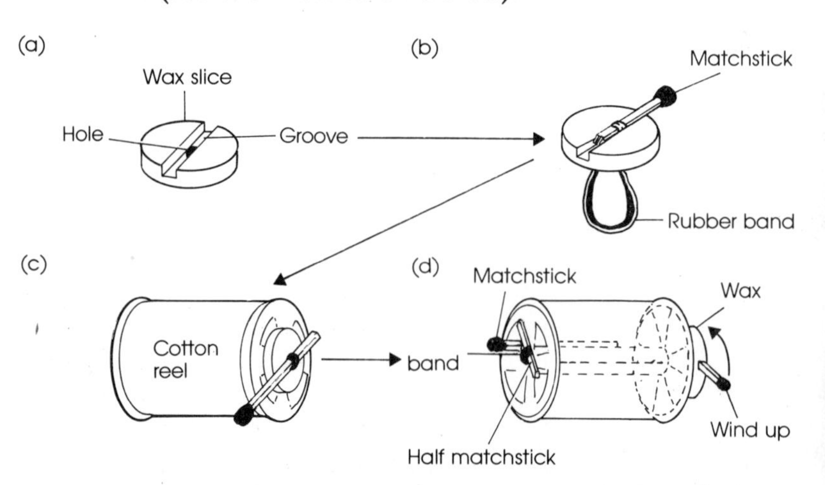

3) Put the rubber band through the hole and slip a matchstick through it.
4) Now pull the other end of the band through the reel and hold it tight with *half* a matchstick.
5) Place another matchstick against this half by pushing it into one of the channels in the cotton reel.

Now wind the tank up by turning the matchstick in the wax groove.

(a) As the rubber band is twisted what form of energy does it gain?

(b) Let the tank go. What form of energy is involved now?

## QUESTIONS ON CHAPTER 2

**1** Supply words or phrases to fill the blanks in the following passage. Do not write on this page.
Energy ____ change energy from one ____ into another. Plants and ____ are living energy converters. The most important man-made energy converters are ____. Two types of ____ ____ engine are the car engine and the ____ engine. Neither of these is anywhere near 100% ____. Most energy is wasted as ____ energy.

**2** What is the *useful* energy change in each of the following energy converters: (a) a car engine, (b) a light bulb, (c) a human being, (d) a plant, (e) a rocket engine?

**3** Copy out the following and fill in the right form of energy in each of the blanks. Do not write on this page.

(a) A coal fire changes ____ energy into ____ energy and ____ energy.

(b) A loudspeaker changes ____ energy into ____ energy.

(c) A burning candle changes ____ energy into ____ energy and ____ energy.

(d) An electric kettle changes ____ energy into ____ energy.

(e) A steam engine changes ____ energy into ____ energy.

**4** Draw a table with these headings:

*Living energy converters* *Non-living energy converters*

Complete the table by giving six examples of each type and drawing a simple sketch of each one.

**5** Write down as many *advantages* as you can think of in using engines. What are the *disadvantages* and drawbacks of engines?

**6** Try to explain the two laws of energy conversion in your own words. Why does the second law explain the so-called 'energy crisis'?

**7** At one time scientists tried to build machines that would carry on moving forever, once they had been started. These endlessly moving machines are called *perpetual motion machines.* Nobody has ever made one. Nobody ever will. Try to explain why. (Hint: look at the two laws on p. 12.)

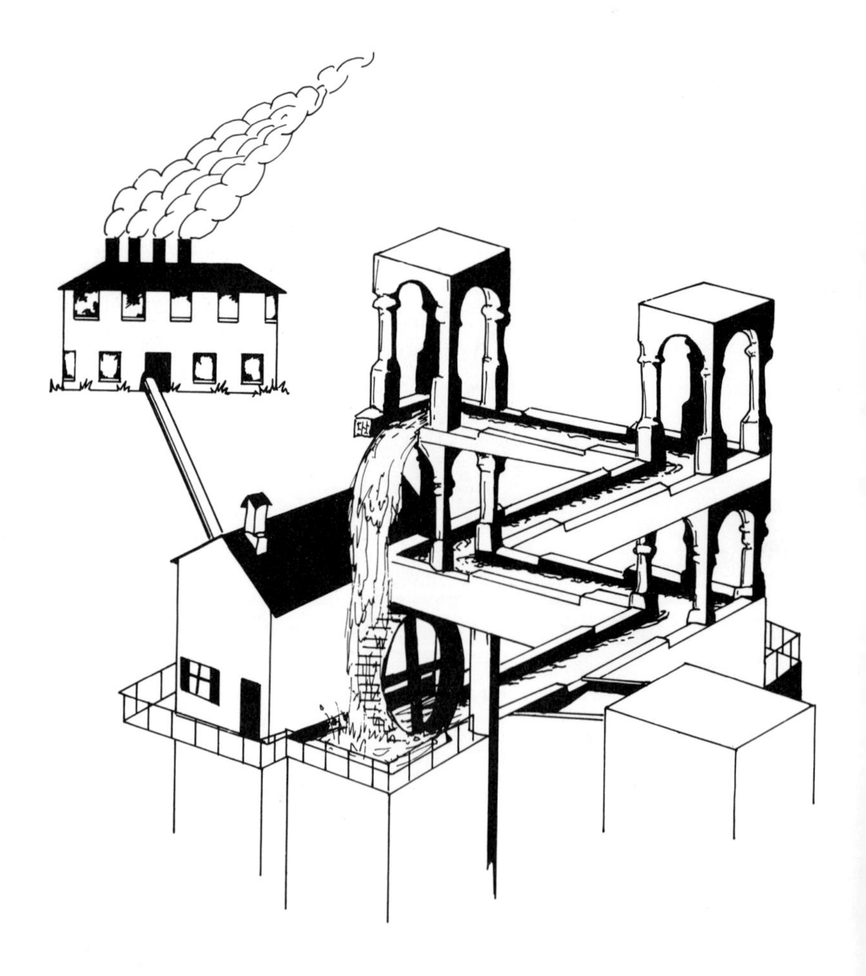

Perpetual energy from perpetual motion. Why won't it work?

# YOU AND YOUR ENERGY SOURCE

## MEASURING ENERGY

You *take in* energy and convert it into other forms. So you are an energy converter, just like plants, other animals and engines. The energy you take in can be accurately measured. This section explains how. But first a short story.

On Christmas Eve in 1818 a boy called James Prescott Joule was born in Salford near Manchester. He studied science and mathematics and, before long, became quite famous. In between his studies he decided to get married. What did he choose for his honeymoon? Bathing in Blackpool? Surfing at Southend? Gambling in Las Vegas? No, none of these. J P Joule measured the temperature of waterfalls. He found that the temperature at the foot of a waterfall was always slightly higher than at the top. Why? He decided that some of the potential energy of the water was converted to heat energy as it fell.

James Joule on his honeymoon

James Joule is now famous for first claiming that heat is a form of energy. His name is given to the unit for measuring energy, *the joule* (J for short).

One joule of energy is a very small amount of energy. So energy is often measured in *kilojoules* (kJ for short). 1 kilojoule = 1000 joules.

## ENERGY YOU TAKE IN

The energy you take in by eating is often measured in kilojoules. For example, teenagers should eat enough food to provide them with between 10 000 kilojoules and 15 000 kilojoules of energy each day. This is their *energy intake.*

Some people will need more than others. It may come from fruit, fish, vegetables, meat, sweets and drinks, or a mixture of these! Here are some examples showing *roughly* how much energy different portions of food may contain:

| | |
|---|---|
| Steak and chips | 3500 kJ |
| Bowl of porridge | 650 kJ |
| Quarter (4 oz) of boiled sweets | 1800 kJ |
| One banana | 500 kJ |
| Glass of milk | 600 kJ |
| Cup of milky tea with one lump of sugar | 200 kJ |
| Glass of lemonade | 400 kJ |
| Pint of beer | 2000 kJ |
| Large fish and chips | 3000 kJ |
| Boiled egg | 400 kJ |
| Slice of bread and butter | 550 kJ |

Remember that the values shown are rough ones. An exact value depends on how big your egg is, the size of your fish and chip portion, the length of the banana and so on.

**ACTIVITY 5**

### What is your daily energy intake?

Imagine a day in your life when you eat and drink only foods from the list above. Make a list of what you might eat, knowing your own appetite. What is your total energy intake for the day, in kilojoules? Is it above 15 000 kilojoules?

## COMPARING ENERGY VALUES IN FOODS

It is rough and inaccurate to compare one portion of food with another. A more accurate way to judge energy values is to compare *100 grams* of one type of food with another. For example, 100 grams of fried liver contains 1020 kilojoules of energy, whereas 100 grams of roast chicken contains only 620 kilojoules. In this way you can compare the energy content of different foods fairly accurately. The figure below shows the energy value of 100 grams of some common foods.

Energy values of different foods (source: Ministry of Agriculture, Fisheries and Food, HMSO)

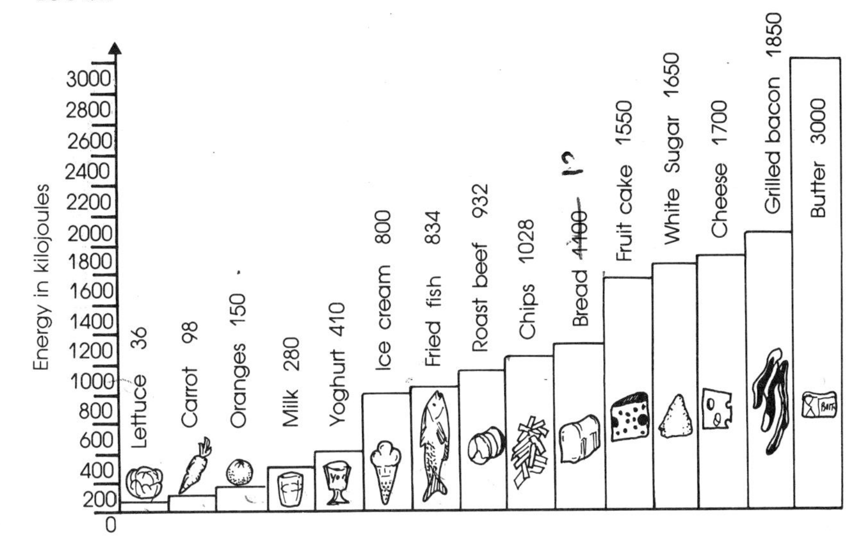


One important fact must be pointed out. In this chapter we are only looking at the *energy* provided by food. There are many other points about the food you eat. Your daily diet should provide you with enough *vitamins* and *protein,* as well as energy. A healthy and balanced diet should contain the important ingredients in the right amounts — and you should not eat too much fat.

## CONVERTING YOUR ENERGY

Energy from food is stored in your body as chemical energy. It is converted when we carry out any activity, e.g. walking, running, reading, or simply breathing. The more active you are, the more energy you need. Different people, depending on their age and size, and their activities, convert different amounts of energy. The figures in the table below show roughly the energy converted in *one day* by different people.

| *Person* | *Energy converted in one day* |
|---|---|
| Lumberjack | 25 000 kilojoules (roughly) |
| Manual labourer | 20 000 kJ |
| Coal face worker | 20 000 kJ |
| School teacher | 12 000 kJ |
| Office worker | 11 000 kJ |
| Teenage boy | 13 000 kJ |
| Teenage girl | 11 000 kJ |
| Baby boy or girl | 4 000 kJ |

As before, these figures are only rough. It all depends on how active the person involved is.

Different activities convert different amounts of energy. The best way of comparing different activities is to compare the energy converted in *one minute.* The graph below shows the energy converted by a 16-year-old in one minute doing various activities.

Energy converted by various activities

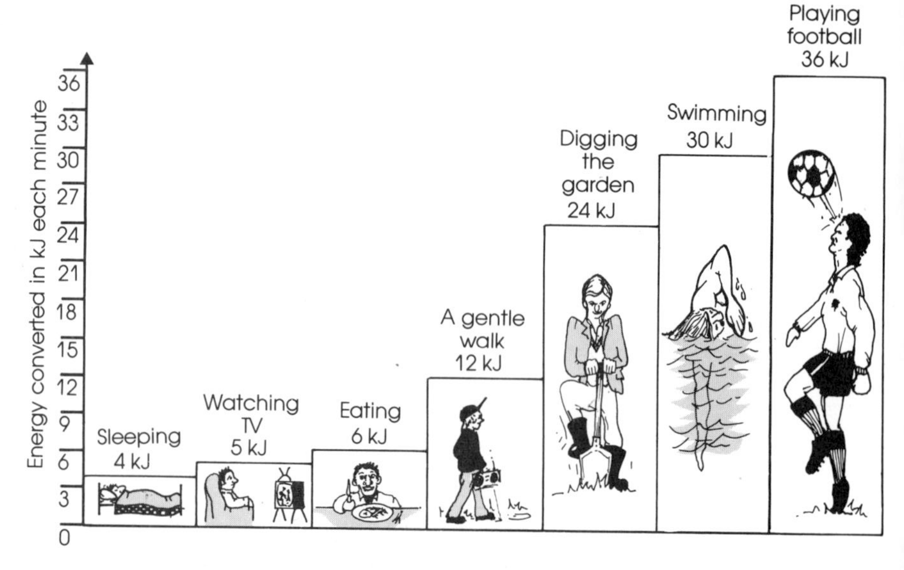


## STRIKING THE RIGHT BALANCE

Human beings are just like all other energy converters. They take in energy (from food) and convert it into other

The human energy balance. Most of the world's population don't have enough to eat

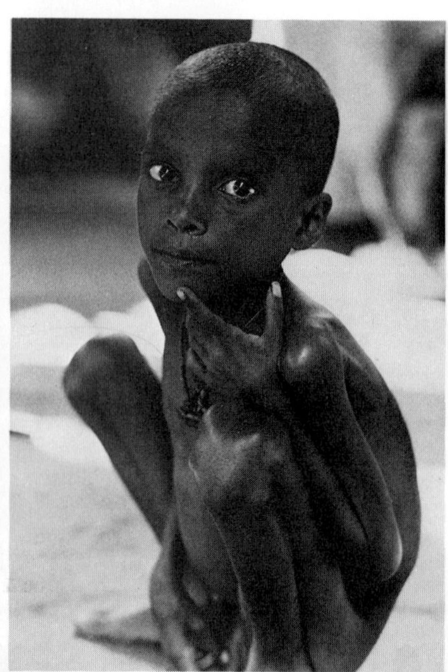

forms of energy by doing various activities. If *more* energy is taken in than a person converts, then he is likely to put on weight. If *less* energy is taken in than a person converts, then he will lose weight. The aim of a balanced diet is to make *energy intake* equal to *energy converted.*

**ACTIVITY 6**

## Burning a peanut

You will need: a sharp pin or needle, a Bunsen burner flame, a test tube or boiling tube, and a thermometer.

1) Pour about 20 $cm^3$ of tap water into the tube.
2) Write down the temperature of the water.
3) Stick the needle into the peanut.
4) Set fire to the nut in your Bunsen flame.
5) Once the peanut is burning hold it under the tube until it has burnt completely.
6) Measure the temperature of the water and write it down.

What has happened to the water in the tube? What is the temperature *difference?* The water has gained energy. Where did this energy come from? What forms of energy are involved? The peanut has been used as a type of *fuel.* Do you think that all foods are fuels?

Burning a peanut

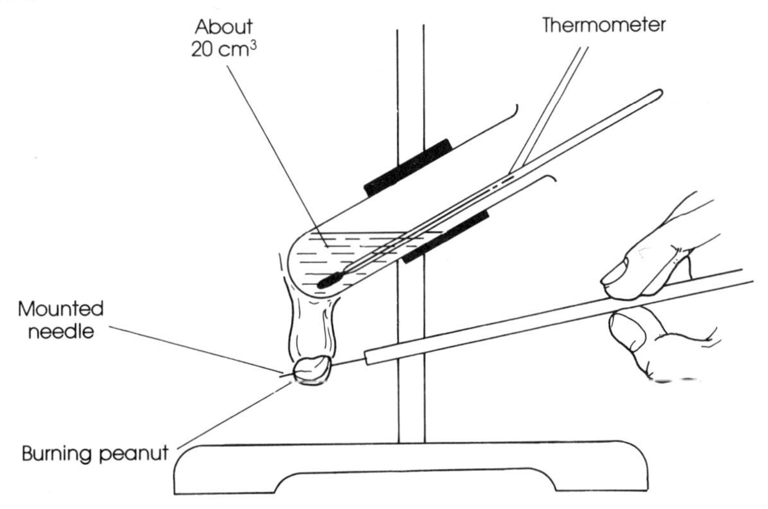


# QUESTIONS ON CHAPTER 3

**1** Supply words or phrases to fill the blanks in the following passage. Do not write on this page.
Energy is measured in ____, named after an Englishman called ____ ____. The energy you take in from food each day can be called your energy ____. For most people it is between ten and fifteen ____ each day. Energy values of foods are compared by measuring the energy in 100 ____ of each food.

**2** Suppose that in *one* day you eat and drink:
a bowl of porridge, a glass of milk, four cups of tea, a plate of steak and chips, a boiled egg and two slices of bread and butter. Work out, roughly, how many kilojoules of energy you take in. How does this compare with the table shown for different people?

**3** Look at the chart on p. 20 showing the energy converted by various activities in one minute. A day lasts for 24 hours, or 1440 minutes. Plan a 24-hour day where you do various activities for a certain length of time (e.g. sleeping, 192 minutes; watching TV, 100 minutes; walking, 30 minutes and so on). Work out how many *kilojoules* of energy you would convert. How does this compare with the number in question 2?

**4** What is meant by a balanced diet?
How much of your energy intake should come from fat? Find out all the ingredients that a balanced diet should have.

**5** Explain why *energy intake* and *energy converted* should be about the same for a human being. What happens if they are *unequal?*

**6** The chart below shows the energy from food eaten by an 'average' person in different countries:

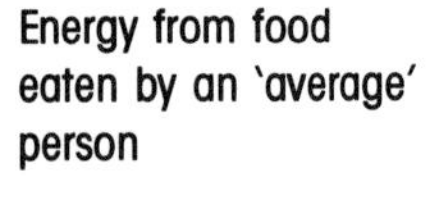
Energy from food eaten by an 'average' person

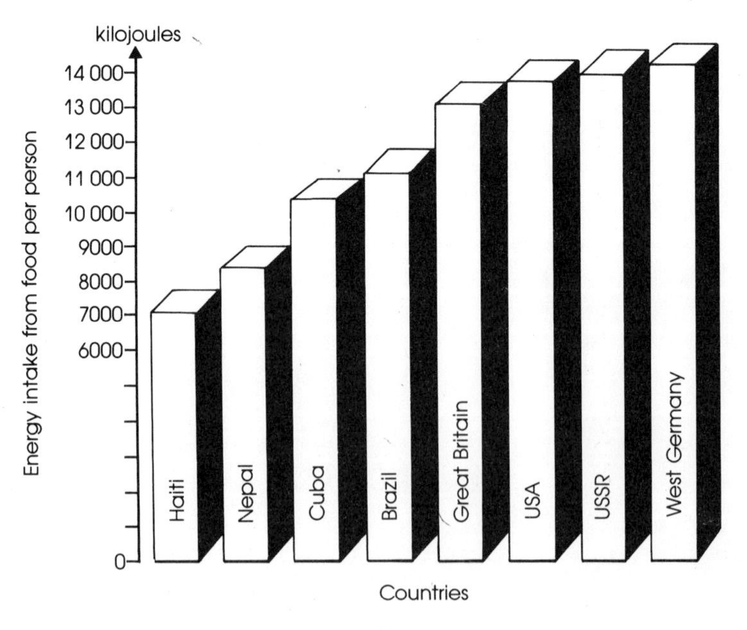


(a) Which country eats the most per head of population?
(b) Which country eats the least?
(c) Why do you think there are such large differences?

CHAPTER 4

# FUELS: ENERGY CAPITAL

## THE FOSSIL FUELS

Gas, coal and oil are called the *fossil fuels.* Why? Because they were all formed from dead animals, trees and smaller plants which lived millions of years ago. Coal was formed when this dead material rotted away and became covered by layers of sand and rock. The dead, rotting plants and animals were squashed down by the rock above them, and after the passage of millions of years they formed a *seam* of coal. The imprints of leaves, other plants and animals were formed on rocks at roughly the same time. These imprints are called *fossils* and they are often found in coal seams.

Mining coal from a seam

Oil was formed between one hundred million and four hundred million years ago from tiny plants and animals which lived in the Earth's vast oceans. When they died they sank to the bottom of the sea and became covered in mud. Layers of mud piled up on top of them. The dead, rotten plants and animals formed into oil and gas over the following few million years!

Oil and gas are often trapped underneath rocks below the sea-bed. They can be reached by drilling through the rock.

Fossil from a coal seam

## WHAT ARE FUELS?

The energy in fuels originally came from the Sun. Plants stored this energy. Animals ate plants and also acted as 'energy stores'. This stored energy in dead plants and animals was trapped and compressed by layers of mud, sand and rock. Fuels were formed. In other words, fuels contain stored energy or *chemical energy.*

Whenever a fuel is used, its stored, chemical energy is converted into another form. This energy conversion is used to do useful jobs, to bring about useful changes.

**ACTIVITY 7**

### Fuels

Here is a list of six different fuels that people use: paraffin, coal, gas, wood, petrol, diesel fuel.

(a) For *each* fuel, write down a useful job that it can do, or a useful change it can bring about.
(b) What forms of energy are involved in each of these jobs?
(c) Where do paraffin, petrol, and diesel fuel come from?
(d) Which one of the six is not a fossil fuel?

## HOW ARE FUELS USED UP?

The fossil fuels took millions of years to form. Yet they are likely to be used up within the space of a few hundred years. Many estimates have been made of how long our fossil fuels will last. Coal looks likely to last longer than the other fossil fuels (350 years?) whereas the world's oil may run dry less than fifty years from now. Gas may run out in 60 years. These estimates assume that people will go on using fuels at the same greedy rate. Where are they being used? Here are some answers:

*Industry.* In Britain most of our fuel is used to provide energy for industry, in factories, offices and workshops.

*Homes.* Britain's homes use at least a quarter of our total energy for heating, lighting and cooking.

*Transport.* Taking people to work, carrying goods around, running buses and trains — all forms of transport need energy. Almost all of this energy comes from *oil,* usually as petrol.

*Farming.* 150 years ago four people out of every five were employed in farming the land. Nowadays, less than two people in a hundred are employed in farming and agriculture. Machines have replaced people. These machines need fuel.

Where do our fuels go?

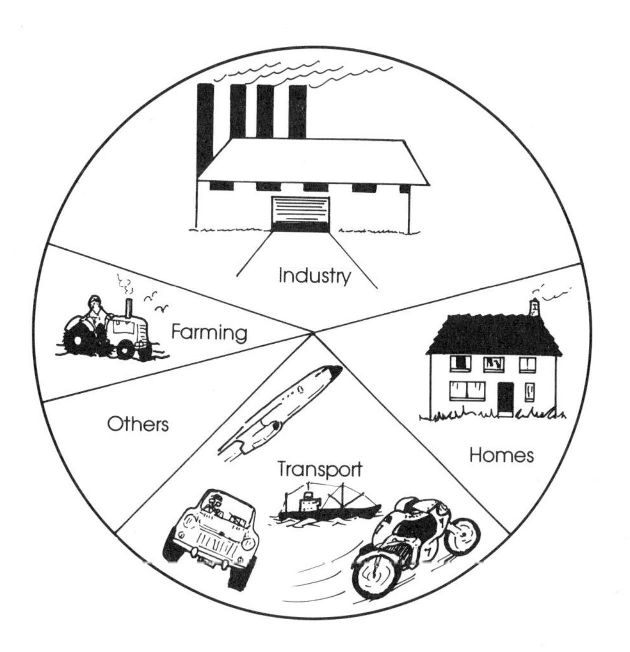


Much of the energy used in industry, homes, transport and farming does not come *directly* from fossil fuels. It comes from *electricity.* At present most of Britain's electricity is generated by burning fuel at power stations. (Electricity is sometimes called a secondary fuel.) Our fossil fuels could be saved and made to last longer if other ways of generating electricity were used. One way is to use nuclear energy, as the next section explains. In later pages, you will see how our energy income will be used more and more in the future to save fossil fuels. But first, another type of energy capital: nuclear energy.

## ENERGY FROM ATOMS

Every substance in the Universe is made up of tiny particles called *atoms.* Over one hundred different atoms are known. They differ because of the way they are made up. Early in this century scientists discovered that every atom has a 'centre', called a nucleus.

### Fission

The nucleus of some of the larger atoms can be broken up, or split. Splitting the nucleus can release tremendous energy. This process is called *nuclear fission.* The energy released by splitting the nucleus is called *nuclear energy.*

Nuclear fission was discovered in 1938 when the nucleus of one type of *uranium* atom (called the U-235 atom) was first split. Seven years later, nuclear fission was used in the first 'atomic bomb' dropped on Hiroshima in Japan. This used a piece of uranium (U-235) slightly larger than a cricket ball. Nearly 1 000 000 people were killed by the energy released.

### Chain reactions

A chain reaction

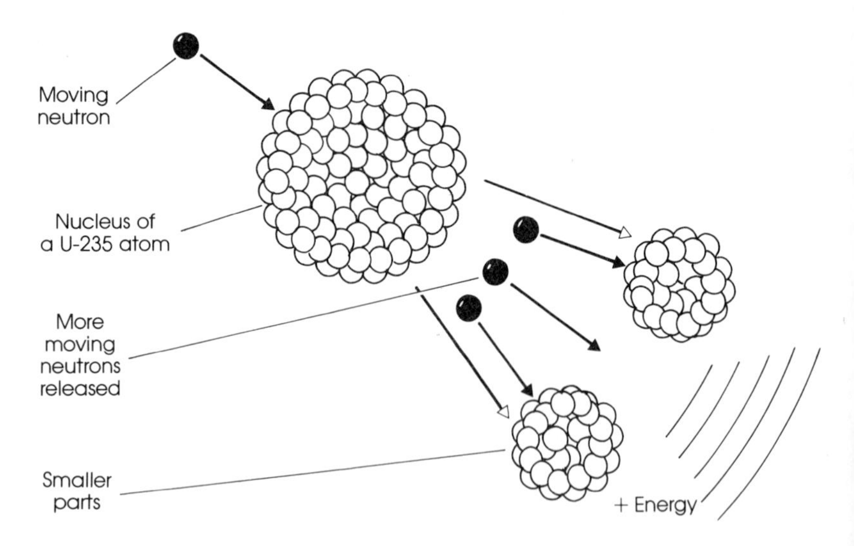


The nuclear energy in an atomic bomb is released in a totally *uncontrolled* way. One uranium atom is split by a tiny particle called a *neutron.* As the first atom splits it releases two or three more neutrons — these go on to split more uranium atoms, which release several more neutrons, which split more atoms and so on. Within a fraction of a second, millions of atoms are splitting almost at once. Tremendous energy is released in the *chain reaction* which occurs.

## Harnessing nuclear energy

Nuclear fission in the atomic bomb is uncontrolled. But it can be carefully harnessed and used by containing the splitting atoms in a large vessel (about the size of a living room) called a *nuclear reactor.* The nuclear energy released inside the reactor can be used to turn water into steam at very high pressure. This steam is used to drive a turbine, which moves a dynamo, which generates electricity.

One type of nuclear reactor called the pressurised water reactor

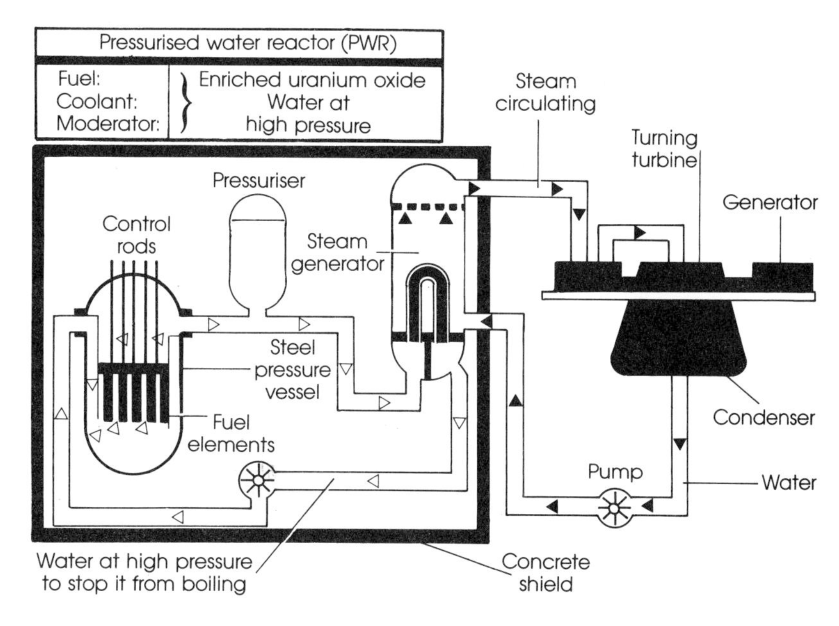


Producing electricity from a nuclear reactor

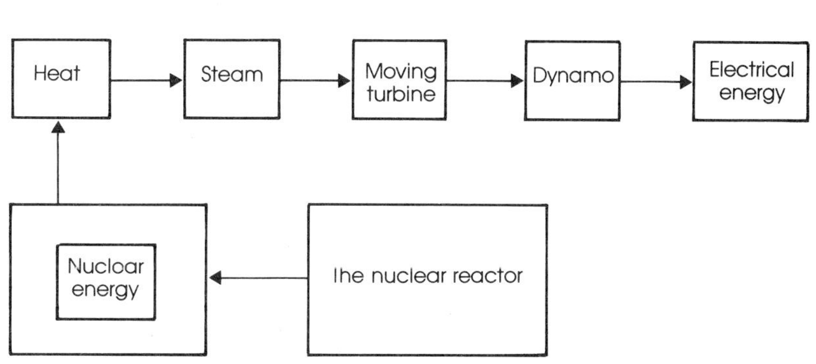


The world's first commercial nuclear reactor started producing electricity in 1956 at Calder Hall in Cumbria. Since then, at least six different types of reactor have been used to generate electricity. One common type, the *pressurised water reactor (PWR),* is shown in the illustration (it is also used to propel nuclear submarines). Every nuclear reactor has the same basic parts:

*Fuel.* The reactor fuel, usually 'fuel rods' containing atoms with a nucleus that can be split, e.g. uranium oxide may be the fuel.

*A moderator.* This slows down the flying neutrons which are produced by fission. In the PWR, water is used.
*Control rods.* These are usually metal rods which can be lowered into the reactor to control the chain reaction inside. They do this by *absorbing* some of the moving neutrons.
*Coolant.* This is used to carry heat away from the reactor. In the PWR, water is used. Without a coolant, the reactor would overheat and even melt.
*Shielding.* The fuel, moderator, control rods and coolant are all surrounded by a huge 'shield'. This shield is made of steel and concrete, several metres thick. It stops dangerous radiation from escaping — flying neutrons and gamma radiation, both of which can kill people.

Nuclear reactors are now widely used in Britain, the USA, Russia and many other countries to produce electrical energy from nuclear energy. But, like the fossil fuels, they have problems and opponents . . .

## PROBLEMS

### Fossil fuels

The obvious problem with coal, gas and oil is that one day they will run out. If we become totally dependent on them, there will be many problems when this happens. Our life-styles are dependent on easy transport, mostly by private car. When the oil wells run dry, how will cars be propelled? Will people be able to adapt their life-styles?

The second problem is pollution. Many of the world's cities are badly polluted, even poisoned, by the fumes from burning petrol in car engines. Burning large amounts of coal is just as harmful. The fumes from Britain's coal-burning power stations put large amounts of sulphur dioxide into the Earth's atmosphere. This is carried by the wind across Norway and Sweden when it dissolves in water vapour in the air and falls to the ground as 'acid rain'. This acid rain has already killed many trees and fish in the forests of Scandinavia. (You can read more about pollution in another *Extending Science* book called *Air.*)

### Nuclear energy

Can nuclear energy avoid the problems of fossil fuels, and even replace them when they run out? Some people believe it can, others say it cannot. Here are the two sides of the argument.

| *Possible problem* | *In favour* | *Against* |
|---|---|---|
| 1 *Safety.* | 'Nuclear reactors are totally safe. No serious accident has happened yet, or ever will.' | 'There is a slight risk that a nuclear accident will occur one day. A nuclear reactor may melt down and dangerous radiation could escape.' |
| 2 *Radioactive waste.* Used fuel and other parts from the reactor are produced as waste products. How can this be disposed of, as it remains radio-active for many years? | 'Radioactive waste from reactors is not a problem. It can be stored safely, or buried underground, or dumped in the sea.' | 'There is no perfect way of disposing of radioactive waste since it remains active for so many hundreds of years.' |
| 3 *Cost.* | 'The cost of electricity from nuclear energy is much less than from other energy sources.' | 'The cost of nuclear energy is rising all the time. It is not the cheapest way to produce electricity.' |

These are just some of the arguments *for* and *against* nuclear energy. A whole book could be written on them. It is often difficult to decide which arguments are facts and which are simply opinions. But one fact is important for nuclear fission. The supply of uranium in the world's mines will one day run out — some people say within 60 years. Uranium mines seem certain to be exhausted before coal mines. Energy from nuclear *fission* may well cover less than a hundred years (1956–????) in mankind's history.

## THE FUTURE OF NUCLEAR ENERGY

### The fast reactor

Millions of pounds are being spent in trying to develop a new type of reactor called the *fast reactor.* This may help to save uranium supplies by actually producing its own fuel. It may even use some of the waste from other nuclear reactors. But no fast reactor is yet producing commercial electricity.

### Nuclear fusion

Another way of releasing nuclear energy is by fusion. This involves forcing very *light* atoms, e.g. of hydrogen, together. The Sun's energy comes from nuclear fusion.

Nuclear fusion

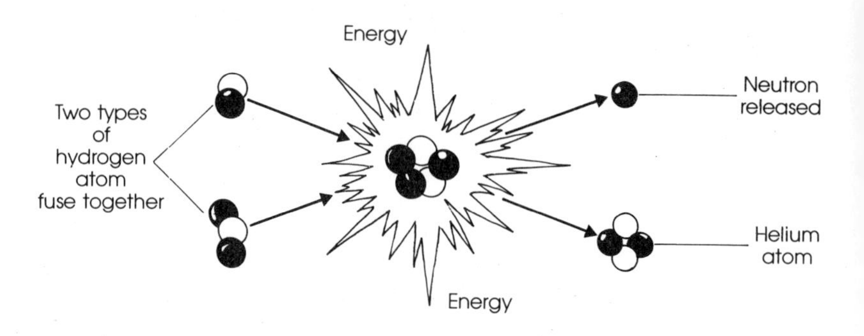


So far, nuclear fusion has only been used in the hydrogen bomb. One day it may be used in *fusion reactors.* These involve trapping the light atoms to be fused inside a magnetic field. The trapped atoms have to be held at about 100 000 000 °C, as hot as the Sun. They form a special kind of gas called a *plasma.* Only then will fusion start to release nuclear energy.

So far, no fusion reactor has been built to produce electricity.

Electricity can be produced by a nuclear fusion reactor

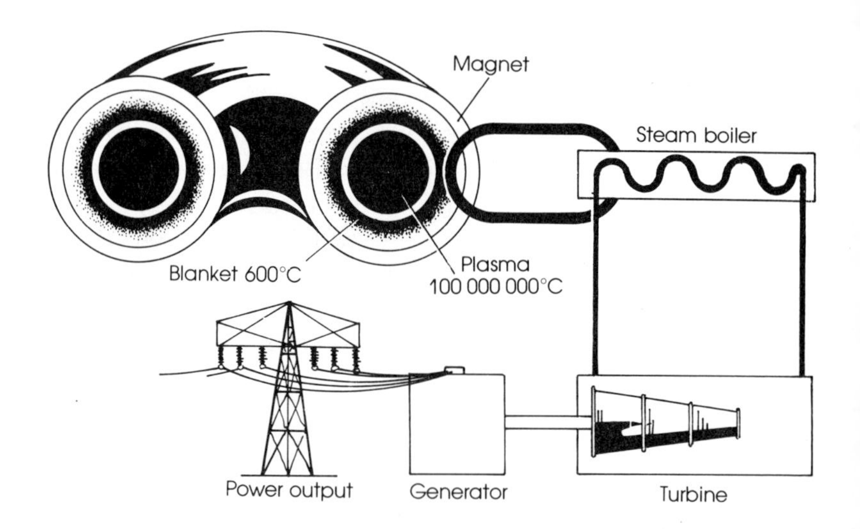


## THE LONGEST GAS PIPE IN THE WORLD

The Soviet Union probably has more gas than any other country. But most of its gas reserves lie in cold, thinly populated areas like Siberia. To carry this gas into more friendly and populated areas, Russia is building the largest pipeline in the world. It will stretch right from Urengoy in Siberia, past Moscow and into West Germany — a total of 3080 miles. The pipeline will cross 500 miles of swamp and more than 300 miles of mountains. Before 1990 it will supply Western Europe with a lot of its natural gas.

## QUESTIONS ON CHAPTER 4

**1** Supply words or phrases to fill the blanks in the following two passages. Do not write on this page.

(a) Gas, coal and ____ are called the ____ fuels. They were all formed from dead ____ and ____ that lived millions of years ago. The energy in these fuels came originally from the ____. Fuels contain stored, ____ energy. This energy is ____ when fuels do useful jobs. Fuels are used by industry, and in homes, for ____ and for ____.

(b) All substances are made up of tiny ____. Large ____ can be split apart to release ____ energy. This splitting process is called ____ ____. Energy from atoms can be used in a nuclear ____. This energy can create steam which turns a ____, to generate ____.

**2** Describe the main parts of a nuclear reactor. What is each part used for?

**3** The pie-charts below show the different amounts of each fossil fuel used in Britain. The first is for 1972, the second is for ten years later in 1982.

(a) Which fossil fuel supplied most energy in both 1972 *and* 1982?

(b) How did the percentage of oil used change over the ten years? Try to suggest a reason for this.

(c) Which fuel was used almost twice as much in 1982 as it was in 1972? Can you suggest a reason?

(d) Make two guesses — draw how the pie-chart might look in 1992; and again in 2002. (Nobody knows the right answer, so you can't get it wrong!)

Amounts of fossil fuel used

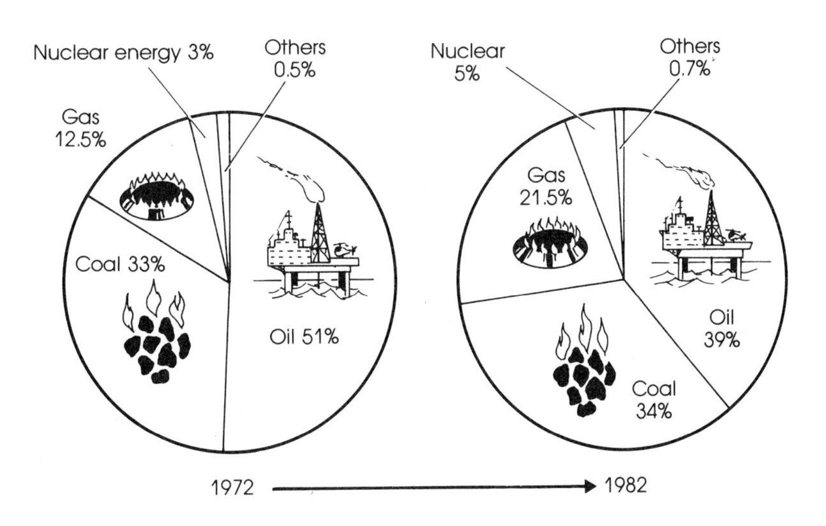

Number of nuclear reactors

**4** This chart shows the number of nuclear reactors owned by some of the world's largest countries in 1981.

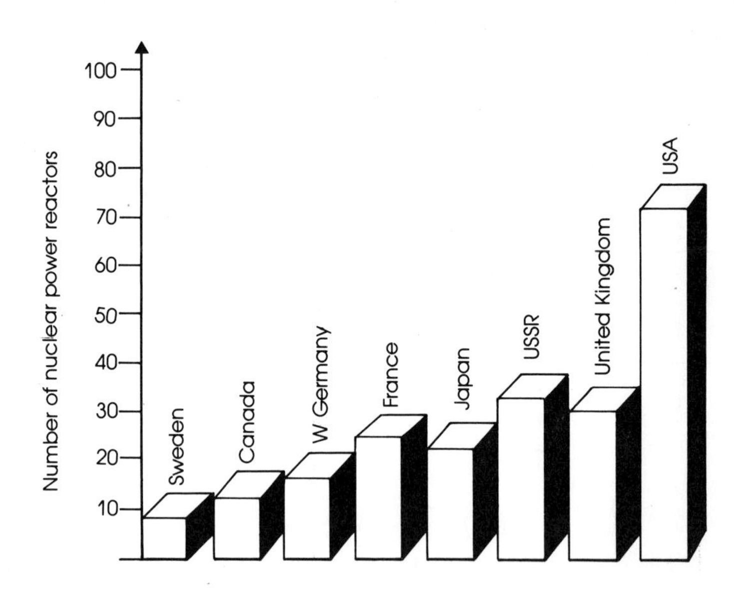


(a) Which country owns the most?
(b) Which country on this chart owns the least?

# RENEWABLE ENERGY SOURCES: ENERGY INCOME

## INCOME AND CAPITAL

Fossil fuels are *not renewable.* Once their chemical energy has been converted to another form it cannot be used again. Some energy sources *are* renewable. The movement energy of the wind and the waves of the sea can be used over and over again. The radiant wave energy from the Sun can bring about useful changes, but is never used up. This table shows some renewable and non-renewable energy sources:

| *Renewable (energy income)* | | *Non-renewable (energy capital)* | |
|---|---|---|---|
| | The Sun's energy (radiant heat and light energy) | Coal<br>Oil<br>Gas | Chemical energy |
| Wind<br>Waves<br>Tides<br>Falling water | Movement energy | Nuclear fuels, e.g. uranium | Nuclear energy |

This chapter explains how renewable sources can be used again and again. Unfortunately, each renewable source has at least one disadvantage.

## DIFFERENT WAYS OF GENERATING ELECTRICITY

Electrical energy is the most convenient form of energy that we have. It can be carried from one place to another, it can be switched on and off, it can be used as an energy source for many different devices.

Ways of generating electricity

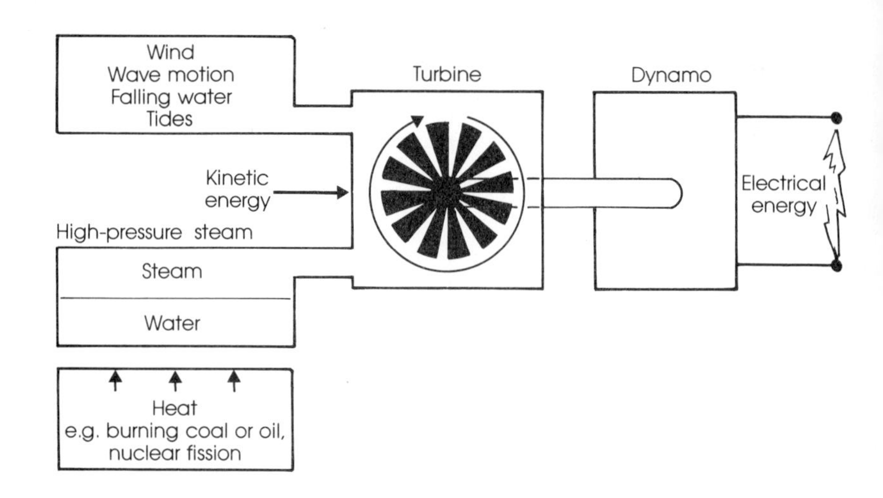


Most of our electricity has been, and still is, generated by burning a fuel (coal or oil), to make steam, which turns a turbine, which drives a dynamo, which generates electricity! The illustration above shows these four steps. But there are other ways:

- Falling water can be used to turn a turbine.
- The wind can be used to turn a windmill, which will turn a dynamo.
- The daily in-and-out movement of tides can be used to turn a turbine and so drive a dynamo.

Four 500 MW turbines in a coal-fired power station near Retford, Nottinghamshire

## Falling water

The kinetic energy of falling water has been used for centuries to turn water wheels. The Romans used water wheels in their flour mills. Nowadays, the kinetic energy of water is mainly used to generate electricity. This is called hydroelectricity. Some hydroelectric schemes use naturally falling water, such as the Niagara Falls. Part of the water is channelled into a pipe. It falls through the pipe and turns a *turbine* at the bottom. As this turbine spins, it makes a huge coil of wire rotate between the poles of a magnet. This generates large amounts of electrical energy.

Other schemes hold the water back with a dam. Some of it is then allowed to rush through a channel, and turn a turbine. Countries with a high rainfall and mountainous ground are ideal for hydroelectricity. In Norway, for example, almost *all* the electrical energy comes from the kinetic energy of falling water. It is a cheap, clean and renewable source of energy.

Bridge and weir at the Rheidol hydroelectric scheme

But there are problems. Building a huge dam can be very expensive. It may also mean flooding a pretty, populated valley with water. Besides this, some countries have neither the rainfall nor the high ground to make hydroelectricity possible.

## Wind energy

Two or three hundred years ago, windmills were used to provide energy all over Britain. Most were used for grinding

corn. But after the steam engine was invented, in 1712, windmills started to disappear. Movement energy was provided by burning coal to make steam, instead of using the energy of the wind.

Now windmills are on the way back. Many people believe that windmills can provide a large part of our energy in the future. Modern wind machines change the kinetic energy of wind into electrical energy. These are called *aerogenerators.* The picture below shows an example. It doesn't look much like the old-fashioned windmill, but it uses the same energy source.

An aerogenerator produces electricity by converting the energy of the wind

As the wind pushes the blades around, the generator spins to make electricity. Small wind generators can provide enough electricity to light a house or to recharge batteries. Large ones can generate up to 2 million watts — enough to run two thousand electric fires or 20 000 light bulbs.

The main advantage of wind energy is that wind blows *most* when we *most* need energy — in the winter. But there are at least three snags in using the wind to produce electrical energy:

- It does not always blow! Alternative supplies will always be needed for 'still' days.
- It sometimes blows *too hard*! Gales, gusts and hurricanes have ruined several wind generators.
- Thousands of wind generators would be needed to provide even part of Britain's energy needs. Windy spots are often in beautiful areas. Wind generators could become an eyesore.

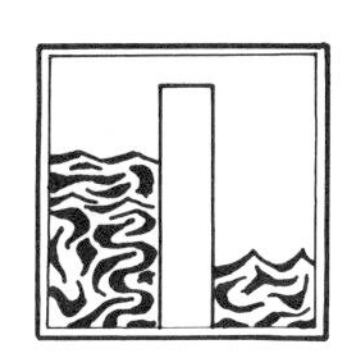

## Energy from the daily tides

Around the coasts of many countries, the rise and fall of the tides can be harnessed to generate a terrific amount of electrical energy. A tidal power station was built in 1966 in France, near St. Malo. This can produce enough electricity to heat and light thousands of homes.

Harnessing tidal power at the La Rance tidal scheme in France

Britain appears to have the perfect place for tidal energy, the Bristol Channel. Schemes for harnessing this energy have been designed. As the tide *rises,* its moving energy could be used to turn a turbine, connected to a generator, to make electricity. Another idea is to trap the incoming tide with a barrage and then hold it back until the tide falls again. At low tide the high water, trapped by the barrage, would be allowed to fall. As it fell it would drive turbines to generate electricity. This is called *ebb generation.* Using this scheme, the tides in the Bristol Channel could generate nearly one-tenth of Britain's electricity supply.

As always, there are drawbacks:

- The barrage, turbines and generators would cost billions of pounds to build.
- The barrage would only generate electricity twice a day.
- Interfering with the tides could affect the birds, fish and other life in the area. These effects are now being studied.

In spite of these drawbacks, tidal energy has been strongly recommended by many scientists. The Bristol Channel has been called 'the best site in the world' for tidal energy.

## Energy from the ocean's waves

The waves on the sea around Britain possess tremendous energy. Some experts believe that this wave energy could supply up to half of Britain's electricity. And, like the wind, *most* energy is available when we most need it — in the winter.

But, as always, there are snags:

- The to-and-fro kinetic energy of the waves has to be harnessed and 'collected' to make electrical energy. One idea is to use floats which bob up and down on the waves. These floats would stretch for miles across the sea near coasts and harbours. They could be dangerous for ships.
- Taking the energy out of the waves could seriously affect the seashores. Some beaches would be without waves. Life in the sea for fish and other animals might be seriously affected.

So far, no 'wave power station' has yet been built. But if certain problems can be overcome, waves may be a valuable energy source for Britain before the 21st century.

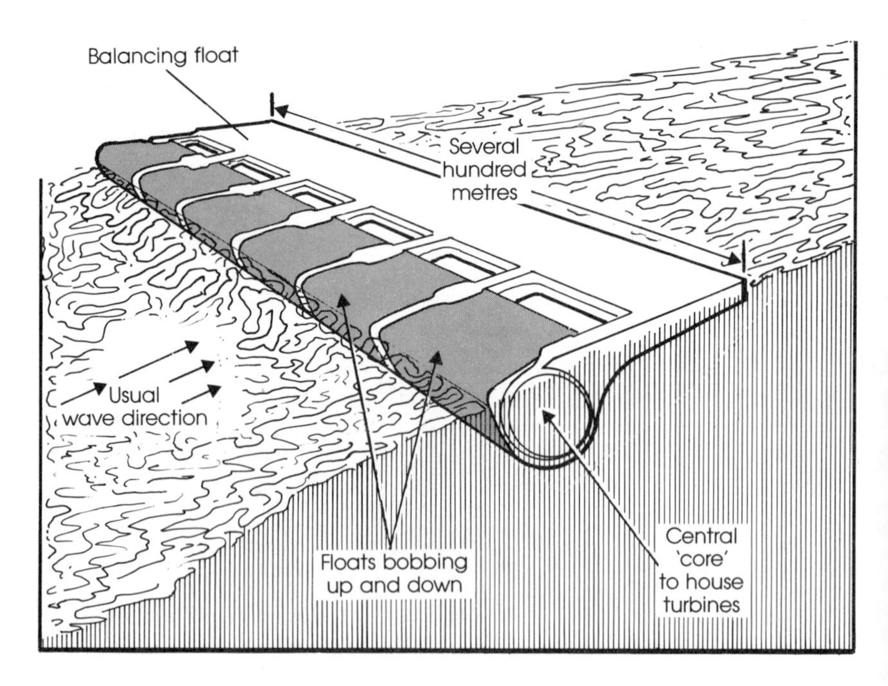


Energy from the waves: one idea — Salter's Ducks

The kinetic energy of falling water, the wind and the daily tides are already being used to generate electrical energy. The ocean waves may be used in the future. But probably the best renewable source of all is the radiant energy that reaches us every day from the Sun.

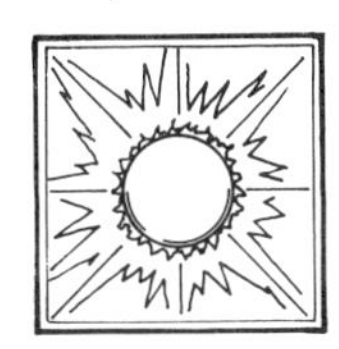

## Solar energy

The radiant heat and light energy from the Sun is often called *solar energy.* The Earth receives more solar energy in *one month* than the total energy in all the fossil fuels we have left. But the problem is, how can we capture and use this solar energy? Here are two of the snags:

*Collecting* the Sun's energy. Large *solar panels* have been built in France and the United States to catch the Sun's energy. But these are expensive to build. Smaller solar panels are already being used in the roofs of people's houses. They can be used to warm their water. But it may take 30 years of use for one of these to pay for itself in Sunny Britain.

*Storing* the Sun's energy. Most of our sunshine comes when we *least* need it — in the summer. The summer Sun's heat could be stored in giant, insulated water tanks until winter. But the storage tank for a house would need to be almost as big as the house itself.

Solar panels in the roof of this house receive and convert the Sun's radiant energy

One idea for the future is the *solar cell.* Solar cells change the Sun's radiant energy directly into electrical energy. But they are expensive. Britain is too cloudy to make them worth using at the moment. Solar cells are often used on satellites in orbit to make the electricity that they need. Some American scientists even thought of collecting the Sun's energy with satellites and beaming it back to Earth.

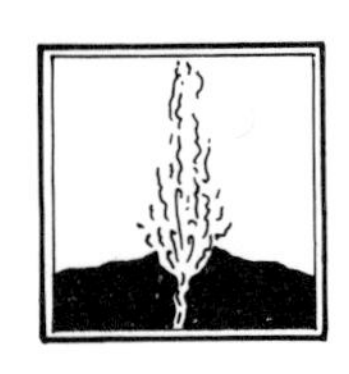

## Energy from inside the Earth

Underneath every part of the Earth there is red-hot, molten rock. Where a volcano erupts, molten rock overflows onto the Earth's surface. In other places water gets down to the hot rock and is sent back up again, hot and steaming. This makes a *geyser* or a *hot spring.*

Hot geyser

Some countries, like New Zealand, already use this *geo-thermal energy,* ('geo', earth; 'thermos', heat, in Greek). One city in Iceland, Reykjavik, heats nearly all its houses from geothermal energy. But a country like Britain has no natural geysers or volcanic areas (though it does have hot springs or *'spas'* at Bath, Cheltenham and Buxton). To use the Earth's heat, artificial geysers have to be made. This can be done by drilling several kilometres down into the hot rock. Cold water is sent down into the hole — it comes back up as hot water, or even steam.

An artificial geyser

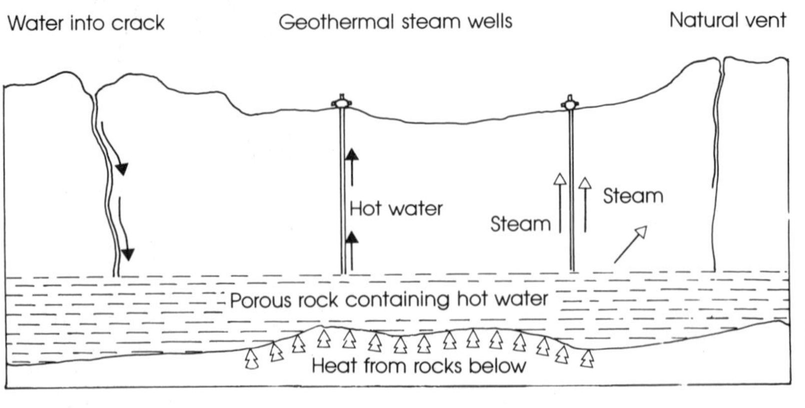


At the moment, this is more expensive than drilling for oil. But oil will run out one day. The Earth's own store of heat may supply a lot of our energy in the next century. And it will never run out.

## Fuel from plants

None of the energy sources mentioned so far can be used as a *fuel.* But almost all of the energy for *transport* comes from one fuel, oil. What will happen when oil runs out? A new fuel will be needed to replace petrol and diesel fuel in engines.

One fuel of the future may be *alcohol.* This may come from the Earth's most important renewable energy source, plants. Plants convert and store light energy from the Sun by photosynthesis. This stored energy can become a fuel. In Brazil, sugar cane is grown especially for making alcohol. These plants are made to *ferment,* rather like hops do in making beer. Then the alcohol is extracted. Brazil hopes to run all its cars on alcohol eventually. (Watch out for drunken drivers!)

Sugar cane is grown in Brazil for making alcohol — a fuel for cars

Another fuel, methane gas, can be made from plant remains (like straw) and animal manure. These can be mixed in special containers called *digesters.* Inside the digester, the plant and animal waste *rot* and make the methane gas. (You sometimes smell this gas when you stir up the mud in a pond.)

Methane can be burnt to heat houses, cook food, generate electricity, or provide fuel for cars. It may be an important fuel for Britain, which has plenty of straw (6 million tons a year) and a lot of cow dung!

ACTIVITY 8

## Boiling water in a paper cup

One of the problems in using the Sun as an energy source is that it shines *most* when we *least* need energy supplies, i.e. on warm days. The answer is to *store* solar energy. You will see that water is an excellent substance for storing heat energy from the Sun. Try this experiment:

Boiling water in a paper cup

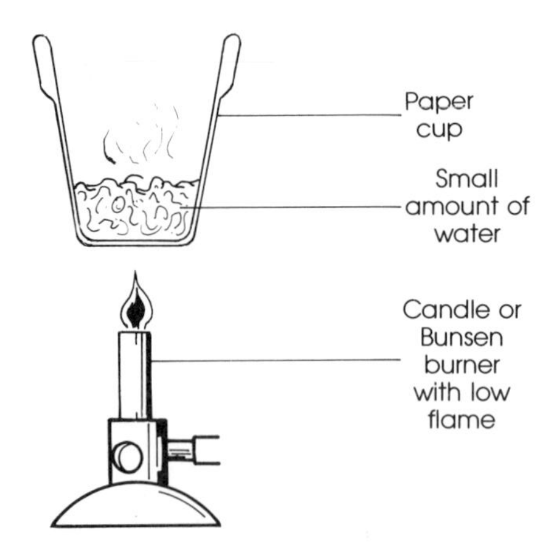


1) Pour a small amount of water into a paper cup.
2) Hold it just above a lighted candle or low-flame Bunsen burner.
3) Make sure the sides of the cup do not catch light.
   (a) What happens after a few minutes?
   (b) Why doesn't the cup burn?
   (c) Where is the heat from the candle?

ACTIVITY 9

## Concentrating and absorbing the Sun's radiant energy

You need: a magnifying glass (a convex lens); two sheets of paper, one white and shiny, the other dull black, and most of all a sunny day.

Hold the lens above the paper until it focuses the Sun's rays onto a small spot on the paper. Do this first for the white paper, then for the dull black paper.

(a) What did you notice? Did either paper start to smoke or even catch fire?
(b) What does this tell you about the difference between white and dull black surfaces?
(c) Which surface would you use to make a solar panel? Why? Why are houses in hot countries often painted white?
(d) What effect does the lens have?

## Sources of energy

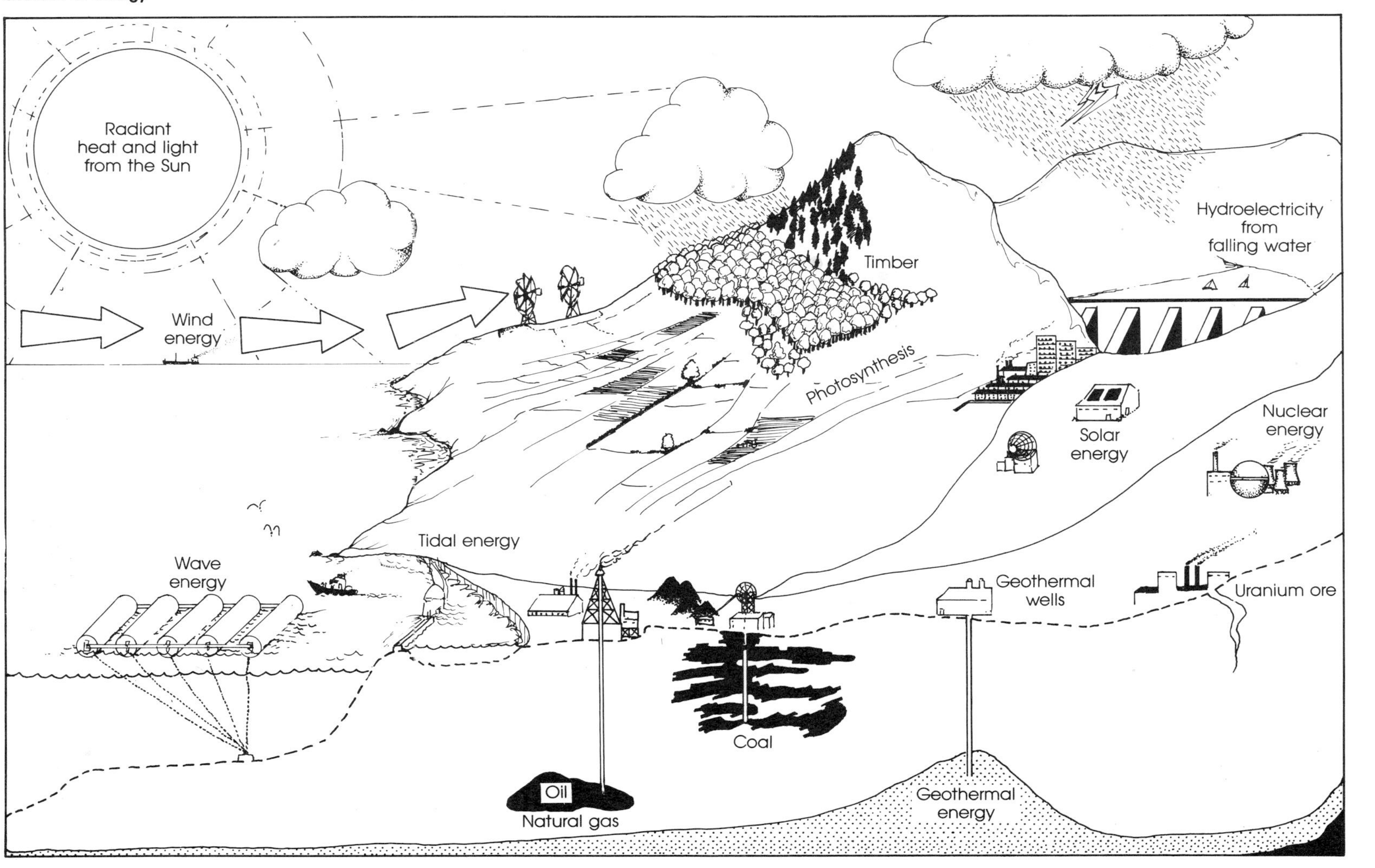

## QUESTIONS ON CHAPTER 5

**1** Make a table showing the *advantages* and *disadvantages* of renewable energy sources. Which renewable energy sources do you think are best for this country? Explain why. Which sources would be best for a continent like Africa?

**2** Geothermal energy is often called a renewable source of energy.
(a) What does the word 'geothermal' mean?
(b) What does the word 'renewable' mean?
(c) Do you think that geothermal energy should be called renewable? Give your reasons.
(d) Find out the names of *three* countries which use geothermal energy. In what obvious ways does geothermal energy show itself?

**3** 'All of the Earth's energy came originally from the Sun.' Is this rule strictly true? If not, give some examples which break the rule. Explain your examples.

**4** The chart below shows the percentage of the electricity used in different countries which comes from hydro-electricity. Study the chart, then answer the questions:

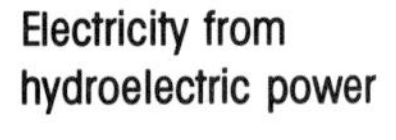
Electricity from hydroelectric power

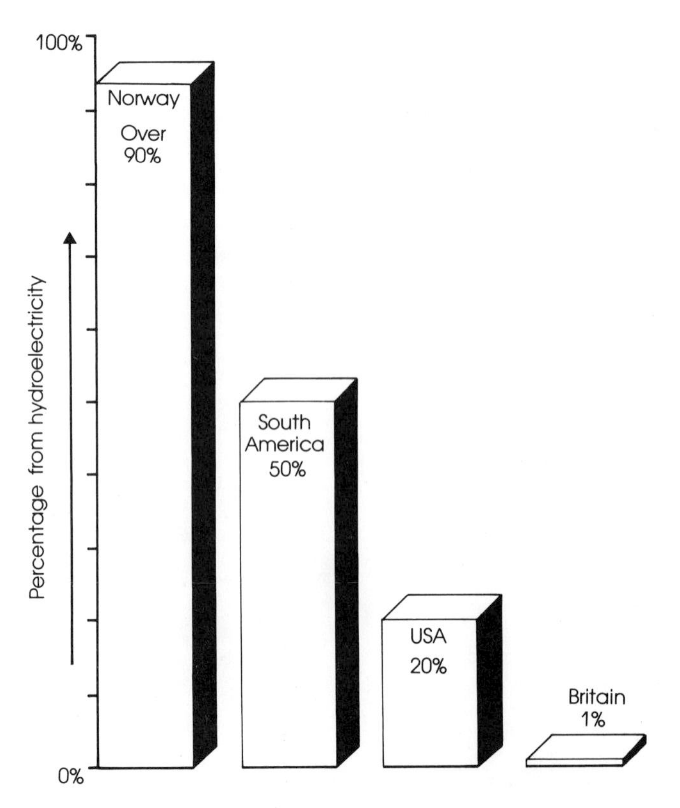

(a) What is hydroelectricity?
(b) Which country shown gets most of its electricity from hydroelectric power stations? Which one gets the least? Can you suggest why?
(c) Why do you think South America is quite well suited to hydroelectricity?

**5** Study the illustration on p. 43 carefully. It shows many of the energy sources which reach the Earth from above *and* below its surface.
(a) Which of these energy sources would you call 'renewable'? Explain why.
(b) Which sources are 'non-renewable'? Why?
(c) Make a table showing:

| *Energy capital* | *Energy income* |
| --- | --- |

Put each of the sources into one column or the other.

**6** This pie-chart shows where the *world's* energy came from in 1980:
(a) Which is the largest energy source?
(b) What sources do you think are involved in 'other'?

The world's energy

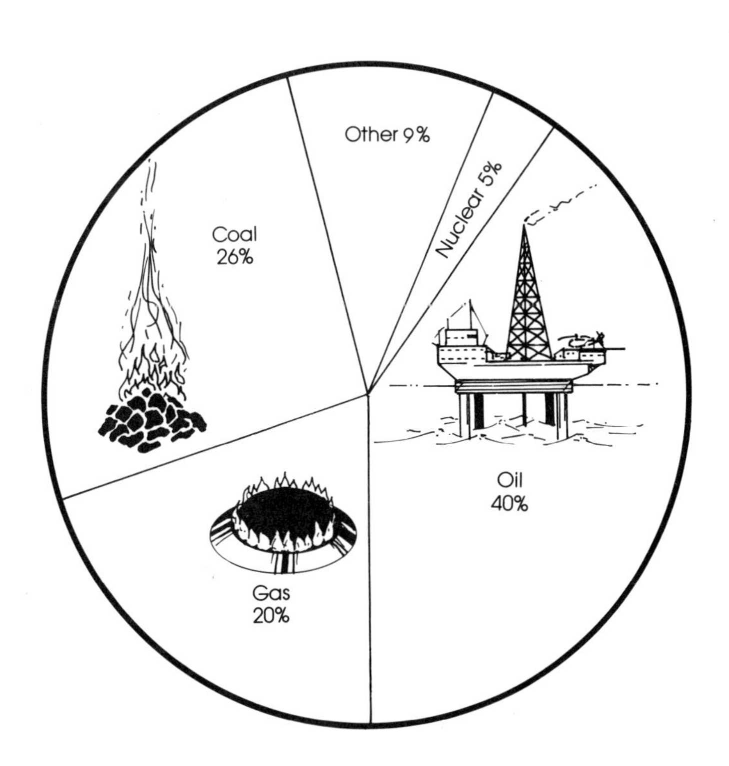

Britain's energy

**7** This pie-chart shows where *Britain's* energy came from in 1980:

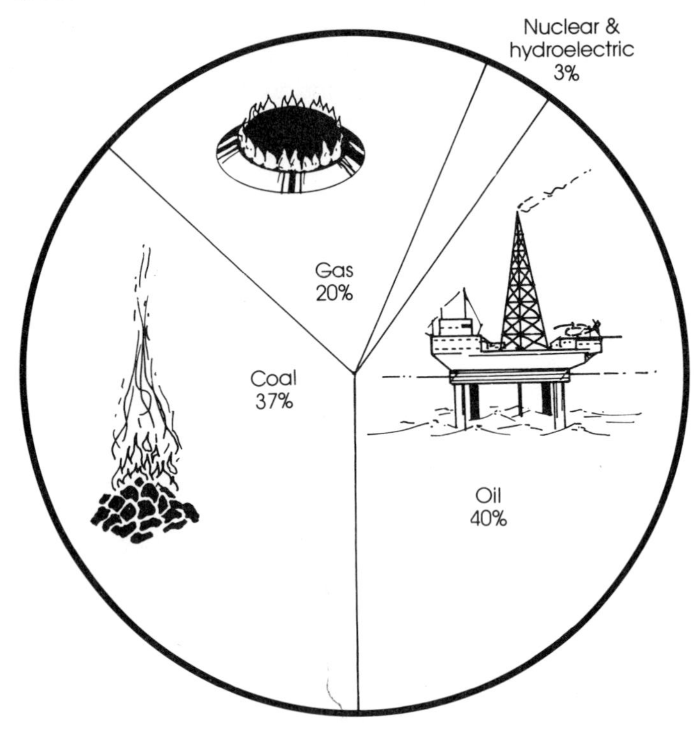


(a) Where do you think most of the oil came from?
(b) Compare this chart with the pie-chart above. Why do you think the proportion of coal is larger?
(c) Can you guess where the hydroelectric energy comes from in Britain?
(d) In 1950, 85% of Britain's energy came from coal. Almost all the rest (nearly 15%) came from oil. Describe how this situation has changed in just over 30 years. Can you explain why?

CHAPTER 6

# MAKING ENDS MEET: ENERGY CONSERVATION

*'Annual income £20, annual expenditure nineteen pounds nineteen shillings and sixpence, result happiness. Annual income £20, annual expenditure twenty pounds and sixpence, result misery'.*

(*Mr Micawber in* David Copperfield *by Charles Dickens*)

## THE ENERGY CRISIS: WHAT IS IT?

Magazines, newspapers and television programmes often talk about the 'energy crisis'. What do they mean? A crisis will occur when the world's people *demand* more energy than the Earth can *supply.* Mr Micawber advised David Copperfield to keep annual *income* greater than annual *expenditure.* In the story, Mr Micawber did not follow his own advice and was put in a debtor's prison.

We have not yet reached Mr Micawber's position with world energy income and expenditure. Indeed, in some ways there is no *energy* crisis. The world has plentiful supplies of energy. As fuels run out people will be forced to harness and use new sources of energy (solar, wind, waves and tides).

But there are two important problems, or crises, that you will meet before the end of the century.

### The food crisis

Many people in the world are already starving. By 1987 the world's population will increase to about five thousand million. This will continue to grow, into the 21st century, perhaps to eight or nine thousand million before it levels off. Food is a source of energy. Growing food, with modern farming methods, requires large supplies of energy. The food crisis could be the biggest problem for the next century.

### The fuel crisis

After the year 2000, the fossil fuels we use now (especially oil) will start to run out. As they do, they will become more and more expensive. People will need to produce *new fuels* before the old ones become too scarce and too expensive.

So there will certainly be a food crisis *and* a fuel crisis in your lifetime. How can these be made less serious? Here are some suggestions:

- by saving energy in many different ways — in heating, cooking, farming and transport (This is called energy conservation.)
- by growing more food, without necessarily using more energy, i.e. by better farming methods
- by finding and developing new sources of energy income and new fuels, before it is too late
- by *sharing* food and energy more evenly throughout the world — some will have less, others more
- by making people more 'energy-conscious' (This will involve educating people, e.g. by reading books about energy!)

The aim of all these suggestions is to *reduce* energy expenditure. The next few sections explain some of these suggestions for energy conservation . . .

## SAVING HEAT ENERGY

Roughly *half* of the energy used in Britain goes towards heating and lighting homes, offices and factories. Eventually, all this energy ends up as useless, low-temperature heat energy, which simply warms up the Earth's atmosphere. Large amounts of fossil fuels and electrical energy could be saved if some of this energy was saved or conserved.

Where heat is lost from a house

Every house loses heat through its walls, windows, doors, roof and chimneys. This heat loss can be reduced in many ways:

*Stopping draughts.* The cheapest and simplest way to stop heat from escaping is to stop cold draughts. Every house needs some draughts for ventilation. But most houses are far too draughty. Strips of foam can be stuck in the gaps around windows and doors. Draught excluders can be fitted underneath outside doors.

*Insulating the loft.* About one-third of the heat from a house escapes through its roof. By placing *insulation* between the rafters in the loft, much of this heat can be saved. In Britain, loft insulation is usually about 100 mm (4 inches) thick. But in colder countries, such as Sweden and Canada, it is 500 mm thick in many houses. Far more of Britain's energy could be saved if thicker loft insulation was used, or recommended by the government.

*Hot water tanks.* In many houses heat is lost from hot water tanks and hot water pipes. Energy can be saved by wrapping an insulating jacket round the tank and foam around hot pipes.

Energy saving in a house: stopping draughts; loft insulation; cavity wall insulation; lagging the hot water tank

*Insulating the walls.* Many houses are built with *two* layers of brick (or concrete blocks) and an air gap between them. This gap is called a *cavity.* Energy can be saved by filling this gap with a special foam which is pumped into the gap through holes drilled in the outside wall.

*Double-glazed windows.* Most houses in Britain have windows made of a *single* sheet of glass. If two sheets are used, with an air gap trapped between them, far less heat escapes through the windows. This is called *double glazing.*

A lot of energy can be saved by double glazing windows. But it is still quite expensive. A cheaper way is to use thick, heavy curtains or better still shutters on the windows. If these are closed at night far less heat is lost.

These are five different ways of saving heat energy in homes, offices and factories. By *conserving* energy in these ways, our fossil fuels could be made to last much longer.

## OTHER WAYS OF SAVING ENERGY

There are many other ways of conserving energy in the future, and avoiding the so-called 'energy crisis'.

*Transport.* About one-fifth of the energy consumed in Britain is used to carry people and goods from one place to another. Most of this is used by road vehicles, i.e. cars and lorries using petrol and diesel fuel.

If people used *public* transport (like buses and trains) instead of private cars for travelling around, a tremendous amount of energy could be saved. A full double-decker bus is about 20 times more efficient at carrying people around in cities than a car with one person in it. Public transport in cities can save huge amounts of energy. But people will not use it unless it is quick, reliable, and cheap — in some cities it is still almost as cheap to go by private car. But as oil and petrol get more expensive, efficient public transport will have to be provided in the future. It would make our oil last a lot longer. There is also the possibility of walking or cycling (more on this in the next chapter).

Is it better by bus?

*Using energy less wastefully.* Our energy supplies are used wastefully. The engines we use are wasteful or *inefficient.* Car engines waste about three-quarters of the chemical energy from petrol supplied to them. Diesel engines are almost as bad. Energy could be saved by making engines and machines more efficient, less wasteful. Engineers are constantly trying to improve the efficiency of engines — but there is a limit to how efficient they can be. Just as much energy could be saved if people used smaller, more economical cars.

*Energy education.* Few people have worried about energy, or the 'energy crisis' until quite recently. The best way of saving energy is to make people *aware* of it! People use energy. People can save it. The next section explains what you and other people can do to save energy.

## WHAT YOU CAN DO

Here are some simple ways of saving energy which can all save *small* amounts. But if 55 million people followed them, thousands of tonnes of oil, coal and gas would be saved each year:

- Turn off unwanted lights. If a room is empty switch off the light.
- Many rooms, especially in offices and schools, are kept too warm. In some cases windows are opened wide while the heating is still on. See if thermostats can be turned down, or radiators switched off before you open windows. Ask the person in charge.
- Different rooms need different temperatures. A sitting-room needs to be warmer than a bedroom or kitchen. Altering heating and thermostats to suit different rooms saves a lot of energy.
- Check that your house or flat is properly insulated. Are you wasting heat through your walls, windows, roof, or doors? Can you reduce the heat loss? Drawing curtains as soon as darkness falls can help in a small way. Keep doors closed. How draughty is your house?
- Don't leave hot-water taps running.
- Don't overfill the electric kettle. You must cover the element, but don't fill it to the top for just one cup.
- Insulate yourself! Retain your 'body energy'.
- Save your empty bottles. Return deposit bottles to the shop — take no-deposit bottles to the nearest bottle bank.
- Beware of 'energy-saving devices'. Often they may save *your* energy, but they all need energy from somewhere

(usually adding to the electricity bill). Petrol-driven and electric lawn mowers are sometimes called 'energy-saving'. Are they really energy-saving devices? Whose energy do they save?

These are a few of the ways in which individuals can be more *aware* of energy, and help to save it. But together, people form a society. The future of a society is partly decided by its *government.* What steps could the government take to stop our energy future from developing into an energy crisis?

## DECIDING OUR ENERGY FUTURE

The first step is to encourage conservation. People can be helped to save energy by giving them money towards loft insulation, putting in solar panels, cavity-wall insulation, or double glazing. Every loft in Britain could be insulated for less than the cost of *one* new power station.

Renewable sources of energy will be around for as long as the Sun 'burns' — at least 40 million years! Money must be spent on harnessing and using these sources in the places best suited to them, e.g. wind energy in high, remote, windy places, wave energy around the coasts, tidal energy in suitable estuaries, solar energy on the South coast, and so on.

More research is needed into nuclear energy. Nuclear fusion could provide endless energy in the next century if only it could be trapped and controlled.

But the most important advice for the energy future is what Mr Micawber might have said:

> *Balanced* energy expenditure and energy income
> — result happiness
> *Less* energy income than energy expenditure
> — result misery

Different societies use energy in different ways. Some societies demand more energy than others. Energy use and energy saving both depend on the society you live in and the way that people live. This is the subject of the final chapter. . .

**ACTIVITY 10**

### Making and testing insulation

You will need: a large beaker, two smaller metal cans, two thermometers, one metal can *larger* than the two small ones, and a clock.

Making and testing insulation: (a) boiling the water, (b) testing your insulation

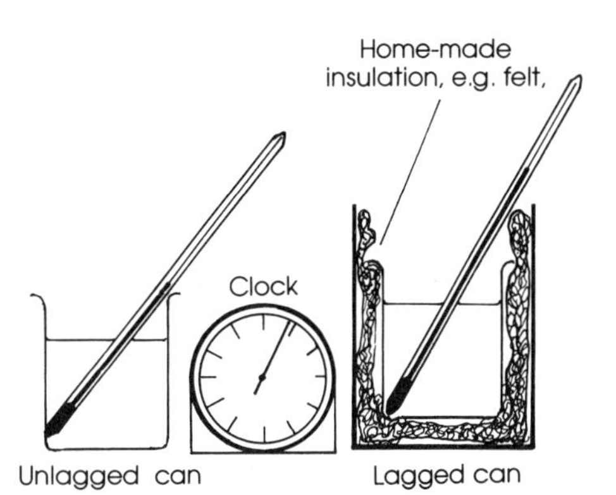


1) Boil about 500 $cm^3$ of water in the large beaker.
2) Place one of the small cans inside the large metal can.
3) Place some insulation (e.g. torn-up newspaper, old felt, fibre glass, cotton wool, paper towels) which you have already prepared between this small can and the large one.
4) Place exactly the same amount of hot water (about 250 $cm^3$) in each of the smaller metal cans.
5) Measure the temperature of the water in each can every two minutes.
6) Draw a table like this and fill it in:

| *Time in minutes* | 0 | 2 | 4 | 6 |
|---|---|---|---|---|
| *Temperature of insulated can in °C* | | | | |
| *Temperature of uninsulated can in °C* | | | | |

What do you notice about your results? Explain them. Try different types of insulation, and compare the results.

Make a wooden or thick cardboard lid. Now try the experiment again using a lid on the lagged can. What difference does a lid make? Can you explain why? Where can lids be useful in saving energy at home?

## QUESTIONS ON CHAPTER 6

**1** What do people mean when they talk about an 'energy crisis'?

**2** Describe *four* different ways of saving energy by reducing the heat lost from a house. Use drawings.

**3** Write down *five* different ways in which you can save energy.

**4** Design and draw a 'perfect' energy-saving house. Explain how each part helps to save energy.

**5** Try to explain why *public* transport (e.g. buses, trains) is less wasteful of energy than *private* transport (e.g. cars).

**6** Look carefully at the chart. It is a forecast of energy supply and demand in the future.

A forecast of Britain's energy demand and supply (from *Energy the Key Resource*, Dept. of Energy)

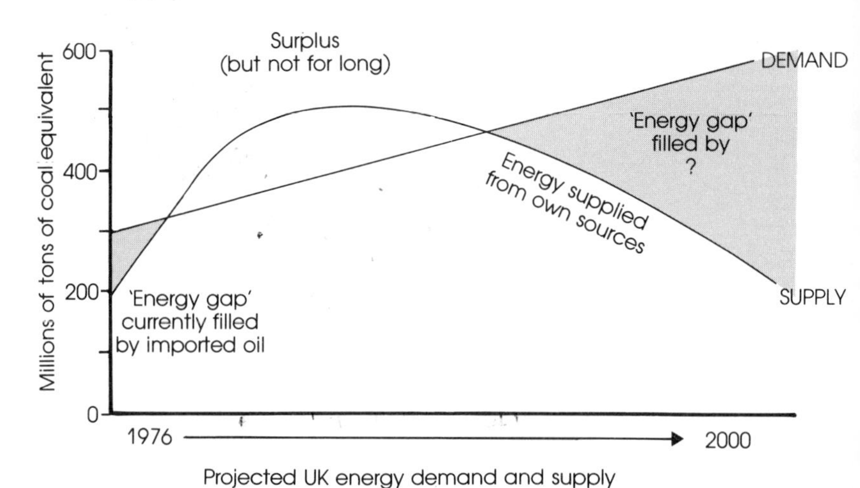


Projected UK energy demand and supply

(a) What is meant by 'energy demand'? What is 'energy supply'?
(b) How was the 'energy gap' on the left of the graph filled?
(c) What is the 'surplus' in the middle of the graph? Explain what *surplus* is. Why do you think Britain had this surplus?
(d) According to this graph, roughly when does the surplus end?
(e) How do you think the future 'energy gap' (on the right of the graph) will be filled? Suggest some ideas for closing this gap by (i) lowering energy demand, (ii) increasing energy supply.

CHAPTER 7

# ENERGY AND DIFFERENT LIFE STYLES

## LIFE STYLES

The previous unit was all about *saving energy* in the kind of society that you live in. In other words, it explained how people can save energy following the *life style* that we are used to now. It is hard to explain exactly what a person's 'life style' is. It involves how you travel to work or school, what you eat, what you do in your spare time, where you go for holidays, what exercise you take, what devices and machines you use, and so on. The life style of people in England is totally different from a person's life style in, say, Africa: different food, different transport, different leisure activities, different work.

What have life styles got to do with *energy*? The answer is: *everything*. The energy used by any society or country depends totally on the life style of its people. People's transport, work, education, defence, eating habits *and* heating habits all affect the energy used, the 'energy expenditure'. At the same time, a society's energy supply and its use affect the way people live. You will see different examples later. But for now remember that in any society:

- energy affects life style
- life style affects energy

## 'QUALITY OF LIFE'

The illustration on p. 22 gives you a rough idea which countries eat the most food in the world as a whole. In some countries it is probably true that many people *overeat*. In most countries a large number of people do not eat enough. This rough 'food guide' to the world's countries also gives you an idea of which countries use the *most energy* for other things. The countries where people eat the most are usually the ones which need the most energy, e.g. the USA, Britain, and the other European countries.

It is clear from the illustration that *energy is not distributed equally* around the world. The way people live in some countries (their life style) demands far more energy than it does in others. This demand for energy depends on a country's industry, its farming methods, its weather, the transport used and the people themselves. Europe and the USA demand large amounts of energy (most from fossil fuels) to support their transport, farming and industry. They are called *industrialised* nations. Nations which use far less energy and have far less industry are often called the *developing* nations.

The important question for people to ask is: how does energy use affect our *quality of life*? 'Quality of life' is another idea which is difficult to explain. The quality of a person's life depends on so many things. The most important needs for a reasonable quality of life are sufficient food and drink, and good health. Once a person has these, education and freedom become important too. An interesting and rewarding job may add to the quality of life. Interests and hobbies, with enough *leisure time* to enjoy them, can also improve the quality of life. It is clear that some countries do not have enough energy, or food, to give their citizens a life style of any quality. This is true in developing (or 'underdeveloped') nations which cannot feed their citizens or offer them education, medicine or even bicycles for transport. At the *other* extreme, perhaps some countries use too much energy. By using too much energy and becoming totally dependent on it for transport, industry and farming, have they gone too far the other way? Have the energy-greedy societies started to *reduce the quality* of life for the people living in them?

Energy use and the quality of life

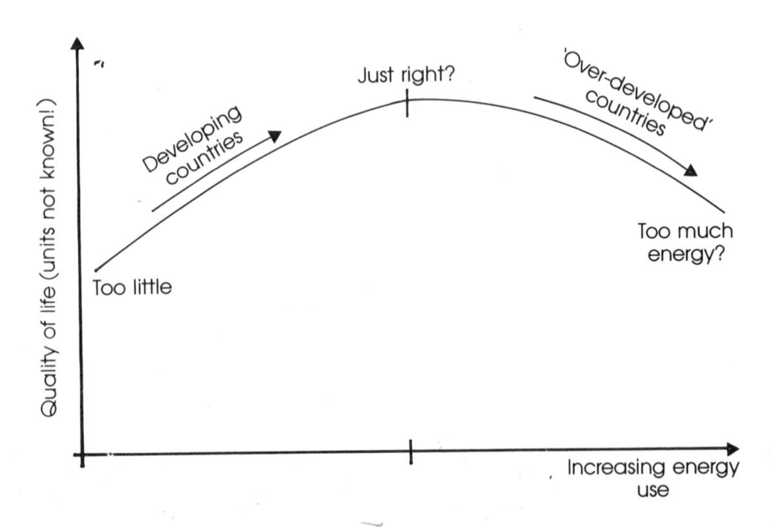

Population size and electricity consumption

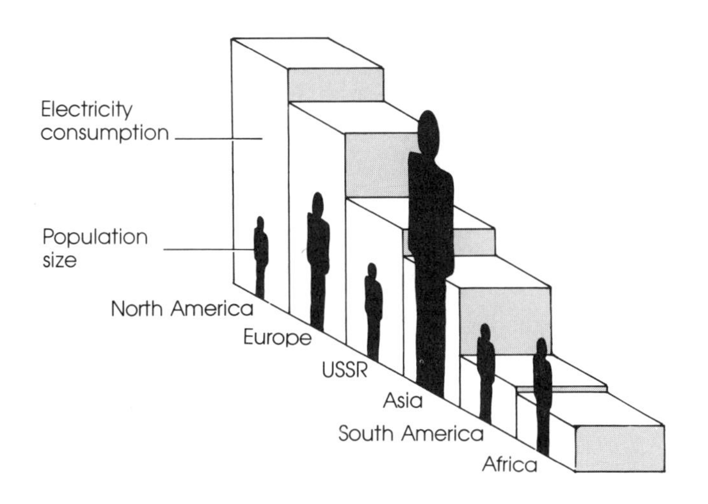


The illustration on p. 56 shows a graph which might reveal how the quality of life for people depends on the energy their society uses. What do you think?

Here are four examples and questions for you to consider:

*Farming.* Modern farming methods are totally dependent on machinery and chemicals — both need large supplies of energy, mostly from fossil fuels. It is sometimes said of modern farmers that 'They convert oil into food'. This is becoming closer to the truth. How does modern farming affect the countryside? Does modern farming reduce or increase your quality of life?

*Industry.* Modern industry depends on large amounts of energy from fuels, rather than from human energy, as it did 100 years ago. Many of the goods made by modern industry are *not built to last.* This is one of the meanings of the 'consumer society' or the 'throwaway society'. Goods are made to last a short time and then be thrown away.

A 'throwaway society' can waste resources when goods are not built to last

This keeps people employed in making them, but is wasteful of energy. Many modern societies depend on short-lived, disposable goods — including cars. Does this add to their quality of life? Does it keep more people employed? Probably not. Thousands of people could be employed in maintaining, renewing and recycling goods of all kinds (from bottles and jars, to cars and lorries) which are intended to last. This would involve more human energy, but would save energy from fossil fuels.

Energy-greedy housing. Council flats in Liverpool

*Housing.* A lot of the housing built in the 1960s and 1970s depended on large inputs of energy. This was the high-rise, factory-built housing which now spoils many modern cities. Many of these tower blocks and huge complexes of flats needed large amounts of energy from fuels, but far less human energy and human labour. Did this modern, energy-greedy housing improve the quality of people's lives? How many were built to last? Some high-rise flats were demolished before they were 20 years old.

*Transport.* Most of Britain's 'transport energy' is used by *road vehicles,* e.g. cars and lorries. Many modern societies are now completely dependent on motor vehicles. Towns, cities, industries and shops have been planned with motor vehicles in mind. Indeed some people believe that motor vehicles have become the *master* not the servant of society. Society is built *around* them, not the other way round. Is the number of motor vehicles, particularly private cars, now reducing the quality of life?

In the 1970s, writer Ivan Illich calculated that the USA puts 45% (almost *half*) of its total energy into motor vehicles: making them, running them and building roads for them.

For transporting people, 250 million Americans consume more fuel than is used by 1300 million Chinese and Indians for *all* purposes. Can a society have *too many* cars for its own good? Ivan Illich estimated that the 'model American' devotes 1600 hours of his time each year to his motor car: washing it, driving in it, parking in it and just sitting in it. He covers a distance of 7500 miles. So his average speed is:

$$\frac{7500 \text{ miles}}{1600 \text{ hours}} \ldots \text{ or } \textit{less than five miles per hour}!$$

He could achieve the same average speed on a bicycle!

In fact, the *bicycle* is probably the best form of transport for the majority of the world's population. The quality of life in many 'underdeveloped' countries could be improved tremendously if each person had a bike. Who needs a car if you can't afford the petrol? In industrialised countries though, it is almost impossible to turn the clock back and encourage people to use bicycles again.

(a) Motor vehicles — servant or master?
(b) Road or rail?
(c) Using the waterways
(d) The bicycle society

(a)

(b)

(c)

(d)

Modern societies are now totally dependent on *motor vehicles.* But there *are* less wasteful ways of carrying people and goods around than the car and lorry . . .

## Carrying people

Most transport uses a *fuel,* usually petrol, to provide its energy. The best way of comparing different kinds of transport is to measure how many *passengers* they carry for how many *miles* using *one* gallon of fuel. For example, you might find that on one gallon of fuel:

- a car will travel 30 miles with two passengers
- a train will travel 1 mile with 500 passengers
- a bus will travel 6 miles with 50 passengers
- a moped will travel 100 miles with one passenger.

How do we compare them? Simply multiply the number of passengers by the number of miles per gallon. This gives the results:

| | |
|---|---|
| car | 60 passenger-miles per gallon |
| train | 500 passenger-miles per gallon |
| bus | 300 passenger-miles per gallon |
| moped | 100 passenger-miles per gallon |

You can see which form of transport carries people most *efficiently.* Different kinds of transport are compared in the illustration below (the estimate for the bicycle is based on the food equivalent of a gallon of fuel, because the fuel used is 'human fuel'!).

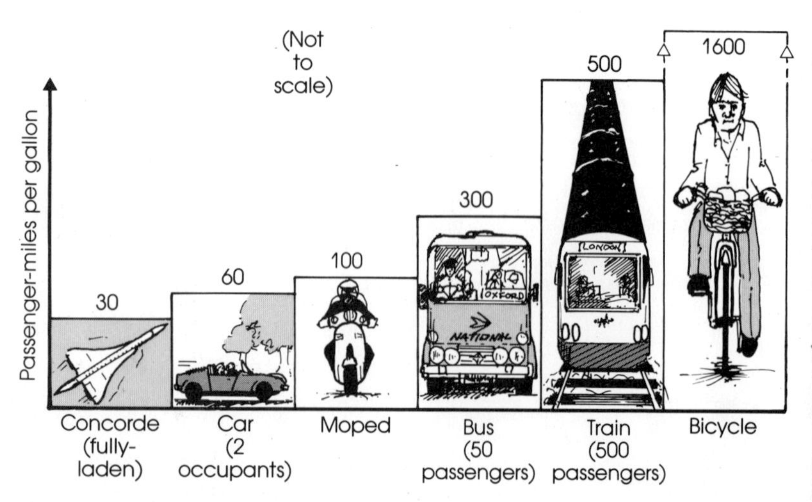


Carrying people efficiently. A comparison

Clearly, the least energy-greedy form of transport is the bicycle. Private cars are a wasteful way of carrying people around. Public transport, by bus or by train, will use far less energy but it needs to be reliable and convenient.

Number of passenger-miles achieved per gallon equivalent by different forms of transport.

Note that the results from e.g. the two trains vary depending on how many people they carry and their fuel efficiency.

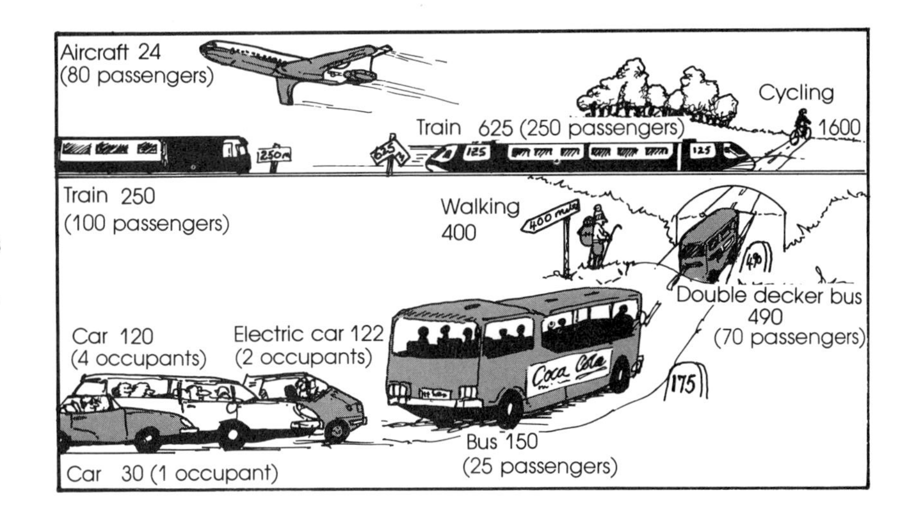


## Carrying goods

You have seen that a society which uses mainly motor cars to carry *people* around is wasteful of energy. Carrying *goods* by road over long distances is also wasteful. Huge amounts of energy could be saved by using railways and waterways, such as canals, to carry more of our goods from one place to another.

Goods transport can be compared by measuring how many *tonnes* of goods can be carried for how many *miles* on one gallon of fuel. For example, you may find that on one gallon of fuel:

- a plane will travel ¼ mile with 12 tonnes of goods
- a lorry will travel 4 miles with 11 tonnes of goods
- a barge will travel 1½ miles with 120 tonnes of goods
- a train will travel 1 mile with 185 tonnes of goods.

Multiplying the tonnage of goods by the number of miles per gallon gives the following results:

| | |
|---|---|
| plane | 3 tonne-miles per gallon |
| lorry | 44 tonne-miles per gallon |
| barge | 180 tonne-miles per gallon |
| train | 185 tonne-miles per gallon. |

The illustration on p. 62 shows the number of tonne-miles per gallon carried by different kinds of transport. You can see that water transport requires less energy than lorry transport to carry goods. Far more use could be made of Britain's waterways (rivers and canals) to save energy.

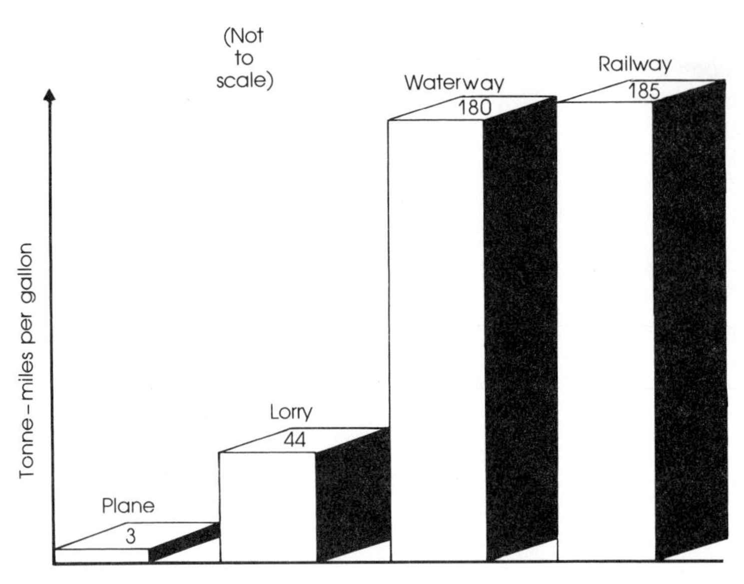


Comparing goods transport

The main point of this chapter has been to explain that the energy needed by any society depends on the life style of the people in it. A society can save huge amounts of energy by changes to its travel and transport, its farming, its industry and its housing. There is one machine which will affect *all* of these things in years to come. This is the computer.

## A NEW AGE?

A device which will affect everyone's life and is *not energy-greedy* is the computer. Indeed, many people feel that the computer may bring a new type of society, 'the post-industrial society'. This new age will *not* involve more and more energy use. The computer, if used sensibly, can reduce our energy needs and even improve the quality of

Computers — the new age? The Johnson Space Center, Houston, Texas

people's lives. Whole books have been written on the way that computers will affect society, save energy and alter your life style. The following are just a few examples:

*Controlling the environment.* Computers can be used to *monitor* ('keep an eye on') heating and lighting in schools, shops, factories and homes. One large supermarket has saved over £1 million on its annual fuel bill by using a computer to monitor and control its lighting, heating and ventilation. In homes, a smaller computer system can be used to control the heat and light in different rooms, or to control energy-saving devices like solar panels.

*Improving communications.* Computers can be used to help people communicate with each other over large distances, perhaps using satellites. Business people can hold 'video conferences', seeing and talking with each other from their own offices. This could save large amounts of energy spent on road and air travel. Also, more and more people may be able to work from home.

*Changing working habits.* As computers and communications improve, people will be able to work and run their own businesses or small industries from home. Microcomputers will be able to communicate with each other and send mail 'electronically' from one computer to another. This may totally change people's working habits. There will be less travelling into and out of big cities, i.e. commuting, to work in large offices. Office work can be done at home. Once again, these changes will save tremendous amounts of energy.

The computer, if used properly, *may* prove to be the best energy-saving device that a modern society can have. It all depends on how the people in that society decide to use it.

## QUESTIONS ON CHAPTER 7

**1** Think about all the different aspects of your own life style. How does it compare with the life style of people in other countries that you know of? Which aspects of your life style require large amounts of energy from fossil fuels? (Consider home, travel, school or work, and your leisure.)

**2** What do you think the phrase 'quality of life' means? Do you think the quality of your life is affected by the energy you use? How? For better or for worse?

3. Consider the quality of life in other countries that you know about. How does it compare with your own? How is this connected with energy supplies — either from fossil fuels or from food?
4. Which type of transport for people is the *least* energy-greedy? Do you think the bicycle could be used in a modern city? What disadvantages does it have?
5. Using waterways is an *efficient* way of carrying goods around. Why do you think Britain's waterways (rivers and canals) are 'under-used'?
6. Explain how computers may reduce energy needs in modern societies in the future.

## ENERGY CROSSWORD

First, trace this grid onto a piece of paper (or photocopy this page — teacher, please see the note at the front of the book). Then fill in the answers. Do not write on this page.

| | | 1 | | | | | | | 2 | |
|---|---|---|---|---|---|---|---|---|---|---|
| | 3 | | | 4 | | | 5 | | | |
| | | | | | | | | | | |
| 6 | | | | | | | | | | |
| | | | | 7 | | 8 | | | | |
| 9 | | | | | | | | | | |
| | | | | 10 | | | | 11 | | |
| 12 | | 13 | | | | | | | | 14 |
| | | | | 15 | | | | | | |
| 16 | | | | | | 17 | | | | |
| | | 18 | | | | | | | | |

### Across

3 The unit of energy is named after him (5)
5 Having 1 down fuels is like having capital in the ____! (4)
6 They use less energy per person than 12 across when they are full (5)
7 It can be driven by steam (7)
9 See 18 across (4)
10 Places where we could use the energy of the waves? (5)
12 They need fuel to go (4)
15 We may run ____ of coal (3)
16 . . . but will energy from the wind come to an ____? (3)
17 They may be heated by 13 down (5)
18 A ____ 9 across uses the Sun's energy (5)

### Down

1 Coal is a ____ fuel (6)
2 They sometimes use 'internal combustion' (7)
4 Insulate your roof to cut down on the energy that is ____ (4)
5 Electrical energy may light it up (4)
6 Ride them to save energy! (8)
8 Nuclear energy is generated in a ____ (7)
11 Water movements to provide future energy? (5)
13 Control ____ may be found in an 8 down (4)
14 A non-renewable energy source (3) . . .
15 . . . and another one (3)

**Solution to Global Energy Wordfinder**

The energy sources are as follows:

| *Renewable* | *Non-renewable* |
| --- | --- |
| Sun | Oil |
| Wind | Coal |
| Tides | Methane |
| Geyser | Gas |
| Niagara Falls | Uranium |

Worrying question: when will the world have an energy crisis?

# INDEX